Additive Manufacturing of Aluminum Alloys and Aluminum Matrix Composites

Additive Manufacturing of Aluminum Alloys and Aluminum Matrix Composites

Editors

Hongze Wang
Mostafa Hassani
Greta Lindwall

MDPI

Basel • Beijing • Wuhan • Barcelona • Belgrade • Novi Sad • Cluj • Manchester

Editors
Hongze Wang
School of Material Science and Engineering
Shanghai Jiao Tong University
Shanghai
China

Mostafa Hassani
Sibley School of Mechanical and Aerospace Engineering
Cornell University
Ithaca
United States

Greta Lindwall
Materials Science and Engineering
KTH Royal Institute of Technology
Stockholm
Sweden

Editorial Office
MDPI
St. Alban-Anlage 66
4052 Basel, Switzerland

This is a reprint of articles from the Special Issue published online in the open access journal *Materials* (ISSN 1996-1944) (available at: www.mdpi.com/journal/materials/special_issues/Addit_Manuf_Alum_Alloy_Alum_Matrix_Compos).

For citation purposes, cite each article independently as indicated on the article page online and as indicated below:

Lastname, A.A.; Lastname, B.B. Article Title. *Journal Name* **Year**, *Volume Number*, Page Range.

ISBN 978-3-7258-0856-4 (Hbk)
ISBN 978-3-7258-0855-7 (PDF)
doi.org/10.3390/books978-3-7258-0855-7

Contents

About the Editors

Hongze Wang

Hongze Wang currently works at the Department of Material Science and Engineering, Shanghai Jiao Tong University, China. Hongze's major research interests include innovative manufacturing processes and systems, particularly in the area of welding and additive manufacturing. His work has made fundamental contributions to the understanding of the relationship among the manufacturing process, the microstructure of the material, and the performance of the product.

Mostafa Hassani

Mostafa Hassani's research interests lie at the intersection of solid mechanics, structural materials, and additive manufacturing. He uses in situ approaches to understand the deformation and failure of structural materials and their underlying physical mechanisms under extreme loading conditions such as impact, high temperatures, small scales, etc. He also leverages extreme conditions of deformation in solid-state to enable novel additive processes for metals, alloys, and composites. Using advanced and operando characterizations, he develops process-microstructure-property relationships for a variety of additive manufacturing and solid-phase processes.

Greta Lindwall

Greta Lindwall is an associate professor at the Department of Materials Science and Engineering at the KTH Royal Institute of Technology in Sweden. Her background is in computational thermodynamics and kinetics and how such computational tools can be developed and applied to alloy development. Dr. Lindwall's current research focuses on materials design for metal additive manufacturing. Her group combines computational approaches with advanced material characterization to increase the understanding and predictability of structure evolution during additive manufacturing. She received her M.S. in engineering physics from LTH Lund University and her PhD in materials science from KTH. Prior to her current appointment, she was a postdoctoral researcher at Penn State's Department of Materials Science and Engineering and at the National Institute of Standards and Technology.

materials

MDPI

Article

Design of Sustainable Aluminium-Based Feedstocks for Composite Extrusion Modelling (CEM)

José L. Aguilar-García [1,*], Eduardo Tabares Lorenzo [1], Antonia Jimenez-Morales [1,2] and Elisa M. Ruíz-Navas [1]

[1] Powder Technology Group (GTP), Materials Science and Engineering Department, Álvaro Alonso Barba Institute (IAAB), Universidad Carlos III de Madrid, Avda. Universidad 30, 28911 Leganés, Spain; etabares@ing.uc3m.es (E.T.L.); toni@ing.uc3m.es (A.J.-M.); emruiz@ing.uc3m.es (E.M.R.-N.)

[2] CIBERINFEC-CIBER de Enfermedades Infecciosas, Instituto de Salud Carlos III, 28029 Madrid, Spain

* Correspondence: josaguil@ing.uc3m.es

Abstract: Additive manufacturing (AM) has become one of the most promising manufacturing techniques in recent years due to the geometric design freedom that this technology offers. The main objective of this study is to explore Composite Extrusion Modelling (CEM) with aluminium as an alternative processing route for aluminium alloys. This process allows for working with pellets that are deposited directly, layer by layer. The aim of the technique is to obtain aluminium alloy samples for industrial applications with high precision, without defects, and which are processed in an environmentally friendly manner. For this purpose, an initial and preliminary study using powder injection moulding (PIM), necessary for the production of samples, has been carried out. The first challenge was the design of a sustainable aluminium-based feedstock. The powder injection moulding technique was used as a first approach to optimise the properties of the feedstock through a combination of water-soluble polymer, polyethyleneglycol (PEG), and cellulose acetate butyrate (CAB) wich produces low CO_2 emissions. To do this, a microstructural characterisation was carried out and the critical solid loading and rheological properties of the feedstocks were studied. Furthermore, the debinding conditions and sintering parameters were adjusted in order to obtain samples with the required density for the following processes and with high geometrical accuracy. In the same way, the printing parameters were optimised for proper material deposition.

Keywords: aluminium alloy; composite extrusion modelling; powder injection moulding (PIM); sustainable feedstock

Citation: Aguilar-García, J.L.; Lorenzo, E.T.; Jimenez-Morales, A.; Ruíz-Navas, E.M. Design of Sustainable Aluminium-Based Feedstocks for Composite Extrusion Modelling (CEM). *Materials* **2024**, *17*, 1093. https://doi.org/10.3390/ma17051093

Academic Editors: Mostafa Hassani, Hongze Wang, Greta Lindwall and Amir Mostafaei

Received: 19 December 2023
Revised: 25 January 2024
Accepted: 26 February 2024
Published: 27 February 2024

1. Introduction

Aluminium alloys are one of the most important materials in the industry, having multiple applications. Their great relevance is due to various properties, such as excellent strength-to-density ratios and recyclability, thermal and electrical conductivity, and low density (~2.7 g/cm^3) [1]. The main sectors where aluminium and, most importantly, aluminium alloys are used are the automotive sector, the construction sector, and the aerospace sector. The most common alloys in aeronautical and automotive components are those of the series 2XXX, i.e., Al-Cu-(Mg) 2024, 2224, 2324, and 2524 (damage tolerance), and 7XXX, i.e., Al-Zn-Mg-(Cu) 7075, 7010, 7055, and 7150 (resistance) hardened by ageing [2]. In aeronautics, the 2XXX series aluminium alloys are used with protective coatings on the underwing or in the fuselage due to properties such as slow crack propagation and fatigue strength. In addition, these kinds of alloys usually have poor resistance to electrolytic corrosion due to the presence of precipitates. For this reason, they are coated with alloys with fewer alloying elements that protect them from corrosion, while these aluminium alloys provide resistance [3]. In the automotive industry, parts made from this type of aluminium alloy are usually manufactured by processes such as extrusion. This method is used to manufacture parts with simple profiles. In comparison, the PIM method used in this study as a support tool has some advantages such as the large-scale production of parts with three-dimensional shapes and greater precision. Furthermore, thanks to CEM 3D-printing technology, which is the technique in which the feedstock designed and optimised in this work will be used for the production of 2024 aluminium alloy parts (such as connecting rods or pistons in the automotive industry [4]), it is possible to produce parts with greater freedom in shape and ability to optimise the design. This allows even greater cost savings in producing these parts, in addition to being able to reuse this feedstock. In this context, CEM technology is a cost-effective process without the need for mass production.

There are different processing methods for aluminium alloys depending on the characteristics or properties required for a specific application or component. Most engineering aluminium alloy parts are obtained by conventional manufacturing processes such as casting, forging, and extrusion. While there are many different methods that attempt to produce final aluminium parts with improved properties, they nevertheless present common drawbacks such as limitations in the design of the parts, difficulty in achieving homogeneity, and high energy consumption. Min Wu et al. [5] performed an isothermal uniaxial compression of sintered porous aluminium alloy 2024 at a semi-solid state. The sintered porous materials were prepared from gas-atomised powders by spark plasma sintering (SPS). This study investigated the deformation behaviour of semi-solid powders and simulated the forming process, achieving high levels of relative density in the final parts. In recent studies, Ning Zhao et al. [6] examined the microstructural evolution and mechanical properties of 6082 aluminium alloy parts produced using a novel forging process that integrates forging and solution heat treatment in one operation. Thanks to this advance, these authors were able to obtain aluminium alloys with mechanical properties that are comparable to those produced by conventional processes, with a homogeneous microstructure. Accordingly, this process was adequate for production, application, and industrial promotion thanks to the cost savings entailed. As a novelty, John Victor Christy et al. [7] investigated recycled aluminium alloy composites produced by Stir and Squeeze Casting. The properties, microstructure, and optimisation of the moulded samples were analysed and showed low porosity and high compressive strength for automotive rims.

More recently, although AM of metallic or ceramic materials is currently focused on the prototyping of samples for subsequent production through other techniques, there is a growing number of areas where 3D-printed samples are being used, including energy storage, aeronautics, medical, and environmental applications. Some of these techniques have been used to produce aluminium alloy parts. Processes such as selective laser melting/sintering (SLS/SLM) can produce samples without specific tooling and shorten the design and production cycle, resulting in significant time and cost savings [8]. Agarwala et al. [9] carried out studies combining the liquid-phase sintering/partial melting (LPS) process with SLS, which incorporates the melting of the lower-melting-point material, thus joining the particles of the structural components of the system. Through these techniques, combined with the use of a deoxidiser or atmosphere control, parts with higher density and less geometric distortion (curling or delamination) were produced. Moreover, Das et al. [10] demonstrated the feasibility of manufacturing components for defence applications using the selective laser sintering/hot isostatic pressing (SLS/HIP) technique. The materials processed by those techniques showed good mechanical properties (hardness and tensile strength) equivalent to conventionally processed materials.

However, the processes described above have some drawbacks, such as difficulty in removing the oxide film from the surface of the metal powder, high energy consumption, and elevated cost. Therefore, this study proposes the use of powder injection moulding (PIM), in particular metal injection moulding (MIM), as a first approach to producing samples, in order to obtain final aluminium components with the highest possible densification by Composite Extrusion Modelling (CEM) [11,12]. PIM is a technology that combines polymer injection moulding and metal powders to obtain metal samples with the geometry of the mould used. Through PIM, it is possible to produce highly complex-shaped components with high reproducibility at a low cost, while obtaining good mechanical properties after the sintering step [13]. In this study, the initial characteristics of the powders (particle size, morphology) have been analysed, taking into account previous studies on obtaining bronze 90/10 components by MIM with different powder characteristics in terms of morphology and particle size. Furthermore, thanks to the appropriate selection of the metallic powder content mixed with the binder system, which provides the necessary fluidity for the correct filling of the mould, it is possible to produce metallic samples by injection moulding [14]. Several feedstocks with different optimal solid loadings were produced by combining metal powders and a sustainable binder composed of polyethyleneglycol (PEG) and cellulose acetate butyrate (CAB), taking into account previous studies on the production of zircon feedstocks with this multicomponent sustainable binder system by powder injection moulding [15,16]. The use of alternative sustainable binders, such as the multicomponent system used in this study, is of great importance in reducing the carbon footprint, making it possible to comply with 2030 Agenda Sustainable Development Goal 12 on responsible and environmentally friendly production (ODS 12: Ensure sustainable consumption and production patterns) [17]. The binder components for aluminium feedstocks are usually combinations of PEG, polymethylmethacrylate (PMMA), and high-density polyethylene (HDPE) [18]. Therefore, the use of the PEG and CAB multicomponent binder system for the design and manufacture of an aluminium feedstock, which has already been tested for other materials such as zircon, as mentioned above, is a great challenge. Also, an in-depth study of the complete process (from feedstock optimisation to moulding, debinding, and final sintering stages) was carried out. The rheological properties of the samples were studied, and the different stages of the process were optimised to achieve materials with high densification. To achieve this purpose, one of the most important points is strict control of the thermal debinding to prevent defects from forming in the internal areas of the material. There are studies on the importance of powder and thermal debinding conditions to obtain the highest possible quality in the final part. Dongguo Lin et al. [19] characterised the rheological and thermal debinding properties of BE Ti64 feedstocks. These authors evaluated the critical solid loading, rheological behaviour, and binder decomposition behaviour, concluding that these rheological and thermal debinding properties of feedstock play an important role in the moulding and debinding stage of powder injection moulding (PIM) and can directly determine the quality of the final PIM product.

On the other hand, in the specific case of aluminium alloys of the 2XXX series, such as the 2024 aluminium alloy used in this study, sintering occurs via a transient liquid phase. These liquid phases migrate along the edges of the particles, penetrate the aluminium oxide layer, and diffuse into the powder particles. As a result of the diffusion of the liquid phase and the re-stacking of the aluminium particles, the porosity of the part is reduced. The success of liquid-phase sintering depends on the ability to control the liquid phase [20]. Thanks to the PIM process as a first approach, it was possible to design a sustainable aluminium-based feedstock in a pellet shape and to optimise the parameters required for its use in the CEM additive manufacturing technology. There are previous studies on the use of the PIM process for the manufacture of MAX-phase feedstocks, which were subsequently used in additive manufacturing technology CEM [21].

Within the broad additive manufacturing technologies, one of the most widely used 3D-printing techniques is Fused Filament Fabrication (FFF). In this technology, a filament is extruded through a nozzle while it is being heated to be deposited layer by layer to obtain the desired geometry. As an alternative to filament processes, Composite Extrusion Modelling (CEM) starts from pellets or granulates. Furthermore, CEM offers some advantages compared to FFF, such as the possibility of using a wider binder percentage/quantity range for 3D printing without the need to use a flexible polymeric binder. In addition, the material can be reused if the printing temperatures are selected correctly and the polymer is not degraded during printing. Therefore, by means of CEM, it is possible to print near-net-shape parts using less aggressive conditions, compared to those required in direct powder bed AM technologies, i.e., laser effects on materials, while also allowing a reduction in cost due to savings on material and recyclability [21,22].

2. Materials and Methods

2.1. Characterisation of Powders and Optimisation of Sintering Cycle

The particle size distribution of the 2024 aluminium alloy pre-alloyed powders (IMR Metal Powder Technologies GmbH, Lind ob Velden, Austria) was determined by laser diffraction using a MasterSizer 2000 (Malvern Instruments, Worcestershire, UK). The powder density was measured with a helium pycnometer (AccuPyc 1330, Micromeritics, Norcross, GA, USA). Additionally, scanning electron microscopy (SEM, TENEO-FEI, Hillsboro, OR, USA) was used to analyse the morphology and microstructure of the cross-section of the powders. To study the thermal behaviour, in order to optimise the sintering behaviour by determining the maximum sintering temperature, the powders were submitted to a thermogravimetric analysis (SETSYS Evolution DTA/DSC, Setaram Instrumentation, Caluire-et-Cuire, France) in a nitrogen atmosphere with a heating rate of 5 °C/min and a final temperature of 800 °C. They were then uniaxially pressed at a low pressure, 340 MPa, in order to reproduce the properties of the brown parts after thermal debinding and before sintering. The green compacts were sintered in a nitrogen atmosphere (1.5 bar) at different temperatures in the interval of 595 to 625 °C for 2 h to design a sintering cycle and to obtain the highest possible densification of the final samples. Scanning electron microscopy and energy-dispersive spectroscopy (EDS) were performed to analyse the microstructure and quality of the sintered samples to determine those similar to the samples obtained after the final sintering stage of the PIM process.

2.2. Production of Feedstocks

For the fabrication of feedstocks with different solid loadings from 57 to 70 vol.%, taking into account similar studies on the manufacture of feedstocks of different aluminium alloys [11,18], the Al2024 powders were mixed with a multicomponent binder composed of polyethyleneglycol (PEG), with two different molecular weights of 4000 and 20,000 g/mol, and cellulose acetate butyrate (CAB), with a molecular weight of 30,000 g/mol, in a Haake Polylab QC (Thermofisher, USA). The composition of the pre-alloyed powder is shown in Table 1, which corresponds to the nominal composition of the 2024 alloy, with high surface purity, according to the supplier (IMR Metal Powder Technologies GmbH, Austria), to maintain a uniform interaction with the binder and promote sintering [11,14,23]. The composition of the multicomponent binder selected for the production of feedstocks is detailed in Table 2. In addition, different additives were used as surfactants (stearic acid, SA) and antioxidants (phenothiazine, PTZ). The importance of this binder system lies in the use of PEG, which is water soluble and eliminates the use of organic solvents, which are generally petroleum derivatives and can be toxic, flammable and even carcinogenic, and CAB, which produces low CO_2 emissions. CAB acts as the backbone in this polymeric binder, and PEG provides fluidity to the mixtures and reduces their viscosity to obtain feedstocks with high solid loadings. Previous studies using zircon feedstocks have shown the good rheological properties of this binder composition for the powder injection moulding process [24]. To study the torque values of different solid loadings in order to propose a critical and optimal solid loading of the feedstock, the same mixer of the mentioned studies was used. For this purpose, the rollers were set to 40 rpm and the torque stabilisation was analysed with a torque rheometer at 180 °C after 0.75 h (45 min) of mixing. In addition, the fluidity of the feedstocks was studied, according to the standard UNE-EN ISO 1133-1:2012 [25]. The test was carried out with a flow meter (CEAST model 6841.000, Pianezza, Italy), in which the different feedstocks were introduced at 180 °C and passed through a hardened steel nozzle with a 2 mm diameter. The weight of the feedstock that went through the nozzle was measured every 0.007 h (25 s). Moreover, the apparent viscosity of the feedstocks was measured in a twin-screw extruder (PolySoft

OS-MiniLab 3, Thermofisher, Waltham, MA, USA) using counter-rotating screws varying the speed from 25 to 250 rpm at 180 °C.

Table 1. Composition of the aluminium alloy.

Element	Si	Fe	Cu	Mn	Mg	Cr	Zn	Ti	Al
wt.%	0.10	0.13	4.67	0.55	1.73	<0.01	0.11	<0.01	balance

Table 2. Composition of the multicomponent binder system used.

Binder	PEG	CAB
Vol.%	70	30
Molecular weight (g/mol)	4000 20,000	30,000
Supplier	Sigma-Aldrich	Eastman

2.3. Stages of PIM Process

Green samples were obtained by injection of the feedstocks, produced as explained above, in a Bimba Flat 1 injector (AB-400, A.B. Machinery, Peotonoe, IL, USA) at 170 °C with an injection time of 6 s and a pressure of 7 bar. To ensure the structural integrity of the samples, and taking into account previous studies with a similar binder system [26], a two-stage debinding process was performed. The first was a debinding cycle in water at 60 °C with continuous agitation and different immersion times (5 h, 7.5 h, and 10 h) for the removal of PEG from the green parts. Next, the samples were dried in an oven for 1 h at 70 °C. After the binder removal with solvents, an interconnected network of pores was generated. This network facilitates the elimination of CAB during the thermal process preventing defects from forming in the samples, and thus CAB can be removed easily without distorting the geometry of the samples. However, the thermal process must be strictly controlled to prevent defects from forming in the internal zones of the material due to an excessive removal ratio of the backbone.

In this context, the dried samples were then placed in a debinding furnace (GD-DC-50, Goceram, Sweden) at 500 °C for 1 h. Although there are studies on the influence of powder characteristics or the debinding cycle, in this work, a specific study was made on the influence of the atmosphere on the removal of the binder. Accordingly, the thermal debinding was carried out under three different atmospheres, including argon (inert, to prevent the oxidation of the parts during heating as much as possible). Non-oxidising N_2 and N_2-H_2 atmospheres were also used for the thermal debinding stage. However, when using an N_2-H_2 atmosphere, hydrides such as AlH_3 and Al_2H_6 are formed. These reactions could cause the partial pressure of H_2 in the pores to decrease and create a hydrogen gradient between the pores and the external atmosphere. This would slow down the filling of the pores since the internal pressure would be higher, as the solubility of hydrogen is high in liquid aluminium. Refilling the pores with hydrogen does not occur when nitrogen is used since it is not sufficiently soluble in liquid aluminium [27]. The heating and cooling rates were 1 °C/min to ensure proper removal of the backbone polymer (CAB). In order to analyse the percentage of binder removal under the different debinding conditions and to be able to choose the parameters that result in the highest percentage of removal, an analysis was carried out of the dimensional and weight changes after immersion of the parts in water and after the drying cycle. The percentage of PEG and CAB removed as a function of time and atmosphere, respectively, was calculated for each condition, following Equation (1):

$$\%Binder\ removed = \frac{m_0 - m_s}{m_{Binder}} \cdot 100, \quad (1)$$

where m_0 is the initial mass of the sample; m_s is the mass of the piece after immersion in water, drying, thermal debinding, and sintering; and m_{Binder} is the total mass of binder in the feedstock according to Equation (2):

$$m_{Binder} = \frac{\%Binder \cdot m_0}{100}, \tag{2}$$

To verify the complete removal of the binder after the two-stage debinding process following the optimal processing variables, a thermogravimetric analysis was performed on an STA 6000 (STA 6000 PerkinElmer, Waltham, MA, USA) in an argon atmosphere with a heating rate of 5 °C/min and a final temperature of 700 °C.

According to the results obtained for the optimisation of the sintering cycle in low-pressure green samples and those of the thermogravimetric study, the brown samples were sintered in a nitrogen atmosphere at 620 °C for 2 h with heating and cooling rates of 5 °C/min.

The relative densities of the pressed and sintered parts were measured. Moreover, the density of the green samples obtained after injection and after sintering was measured by the Archimedes method to obtain the relative density and the closed porosity, also analysed with ImageJ (Fiji, version 2.35) software.

3. Results and Discussion

3.1. Characterisation of Powders and Determination of the Sinterability

As explained in the previous sections, both the morphological features and the chemical composition of powders must be suitable for the injection process and the subsequent sintering. In this sense, an SEM analysis of the Al2024 aluminium alloy powder particles was carried out. The results can be seen in Figure 1, where the spherical morphology of the particles can be observed. This spherical morphology provides better flowability (the feedstock solid loading can be increased and presents less shrinkage in sintering after injection) [11]. In addition, they are pre-alloyed alloy powders, which favours the homogeneity of the alloy. During the sintering stage, the diffusion process of the alloying elements, the formation of sintering necks, and the liquid phase are more favoured as these elements are more homogeneously distributed within the particles. Therefore, and as will be explained later, this diffusion of alloying elements occurs from the centre of the particles towards the particle boundaries, where the liquid phase is generated.

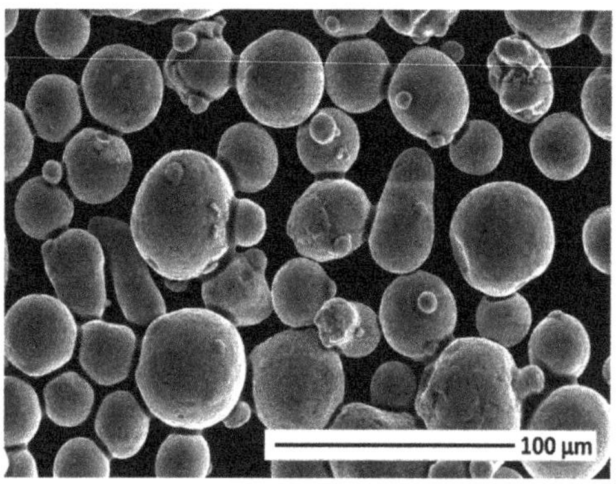

Figure 1. SEM micrograph of the Al2024 alloy powders.

On the other hand, the particle size distribution of the powders was determined with the MasterSizer 2000 analyser. This analysis showed a wide particle size distribution (Figure 2 and Table 3), ranging from 28.89 µm (D_{10}) to 82.88 µm (D_{90}). A wide particle size distribution of the powders indicates that they are optimal for injection moulding, as they will have a low viscosity and are easy to mould.

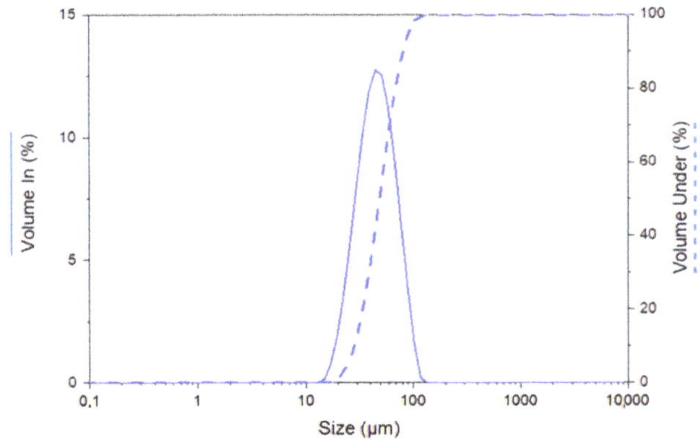

Figure 2. Particle size distribution of the Al2024 powders.

Table 3. Mean particle sizes of Al2024 alloy powders, obtained from particle size distribution analysis by laser.

D_{10} (μm)	D_{50} (μm)	D_{90} (μm)
28.89 ± 0.01	49.85 ± 0.02	82.88 ± 0.03

In order to perform the final sintering of the parts after thermal debinding, it is important to analyse the thermal behaviour, with the melting temperature of the feedstock metal-based material, among others. A DTA-TG analysis was carried out with the 2024 aluminium alloy powders to obtain the heat flow curve of this material (Figure 3) and thus be able to establish an optimal sintering temperature. In this test, it is evident that there are two endothermic peaks, a small one at 509 °C and a big one at 647 °C, corresponding to the melting point of this pre-alloyed powder with the onset temperature at 635 °C. The occurrence of a phase transition during heating was suggested by the endothermic peak at 509 °C, which corresponded to the production of a liquid phase in aluminium by a eutectic reaction. Therefore, the solidus temperature of the powders of this 2024 aluminium alloy was 509 °C, consistent with analyses from previous studies of this alloy [28]. In addition, a mass gain can be seen at 500 °C due to oxidation of the aluminium powders.

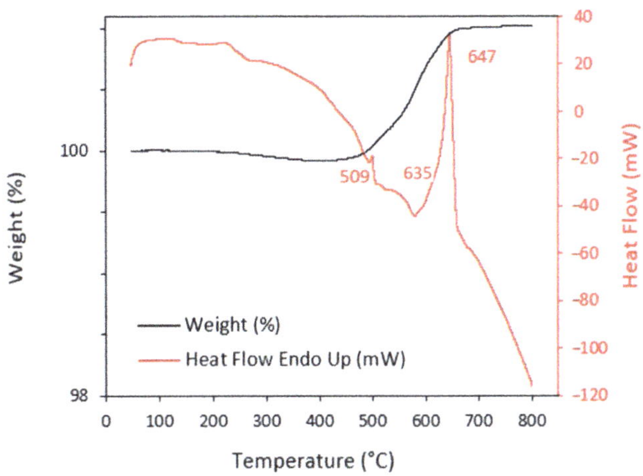

Figure 3. DTA-TG analysis of the Al2024 alloy powders.

Considering the results shown above regarding the thermal behaviour of the powders, and to further examine the sinterability of samples after debinding (brown samples), the powders were uniaxially pressed at low pressure (340 MPa) and sintered in a nitrogen atmosphere (1.5 bar) at a temperature range of 595–625 °C for 2 h. The theoretical density of the 2024 aluminium alloy, with a value of 2.8 g/cm^3, was obtained by the measurement of the powders' densities in a helium pycnometer. Green and sintered densities were measured by the Archimedes method. The relative green density of these low-pressed powders was 84.2% (Figure 4, blue dotted line) and that of sintered samples at different temperatures are shown in Figure 4.

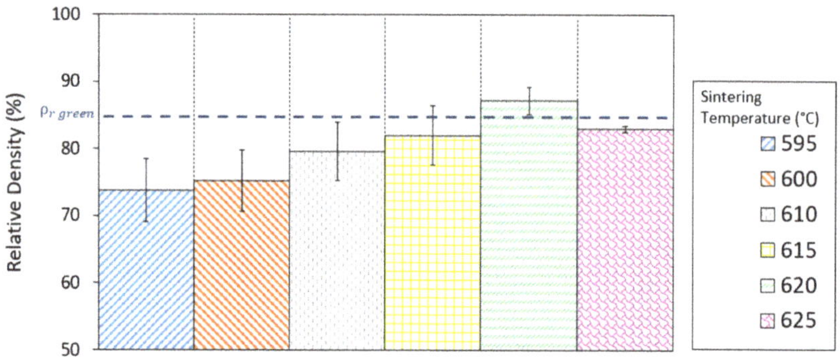

Figure 4. Relative density of Al2024 parts sintered at different temperatures from 595 to 625 °C.

Density values obtained for the pressed and sintered parts were markedly low compared to the theoretical density of the 2024 aluminium alloy, due to the presence of a high level of porosity (see Figure 5) and swelling of the parts. However, it can be observed that the only parts that densified by obtaining a relative density higher than the green density were those sintered at 620 °C, with a relative density value of 87.2%. These parts suffered the least distortion and shrank (2.39 ± 0.19% on the X axis and 1.31 ± 0.31% on the Y axis). Because of the need to improve this, some studies have focused on the benefit derived from the addition of small amounts of tin to the composition [29,30]. As can be seen, the highest relative density after sintering was obtained with the sintering temperature of 620 °C. Density decreased again in samples when the sintering temperature was increased to 625 °C, showing higher porosity. This indicates that the optimal sintering temperature was overcome.

SEM analysis of the cross-sections of the Al2024 samples after sintering at 595, 600, 610, 615, 620, and 625 °C was carried out to analyse the sintering process and the porosity of the samples (Figure 5).

As a pre-alloyed aluminium alloy powder, a supersolidus sintering process was produced. The relative density increased with increasing sintering temperature, as can be seen in the micrographs obtained, which show a decrease in porosity with increasing temperature. It should be noted that this pre-alloyed aluminium powder is always covered by a thin oxide layer (Al_2O_3). This implies that the liquid phase generated during sintering is prevented from penetrating through this oxide layer. Therefore, in samples sintered at low temperatures, the formation of the liquid phase was not sufficient to achieve high densification, and porosity was generated. However, at the sintering temperature of 620 °C, a network of the former liquid phase was observed. This liquid phase was generated from the Cu- and Mg-rich precipitates in the particle boundaries, as verified by subsequent EDS analyses. These eutectic phases promoted the densification and re-stacking of the particles and a remarkable decrease in porosity, as in other studies on the sintering of aluminium powders [28,31,32].

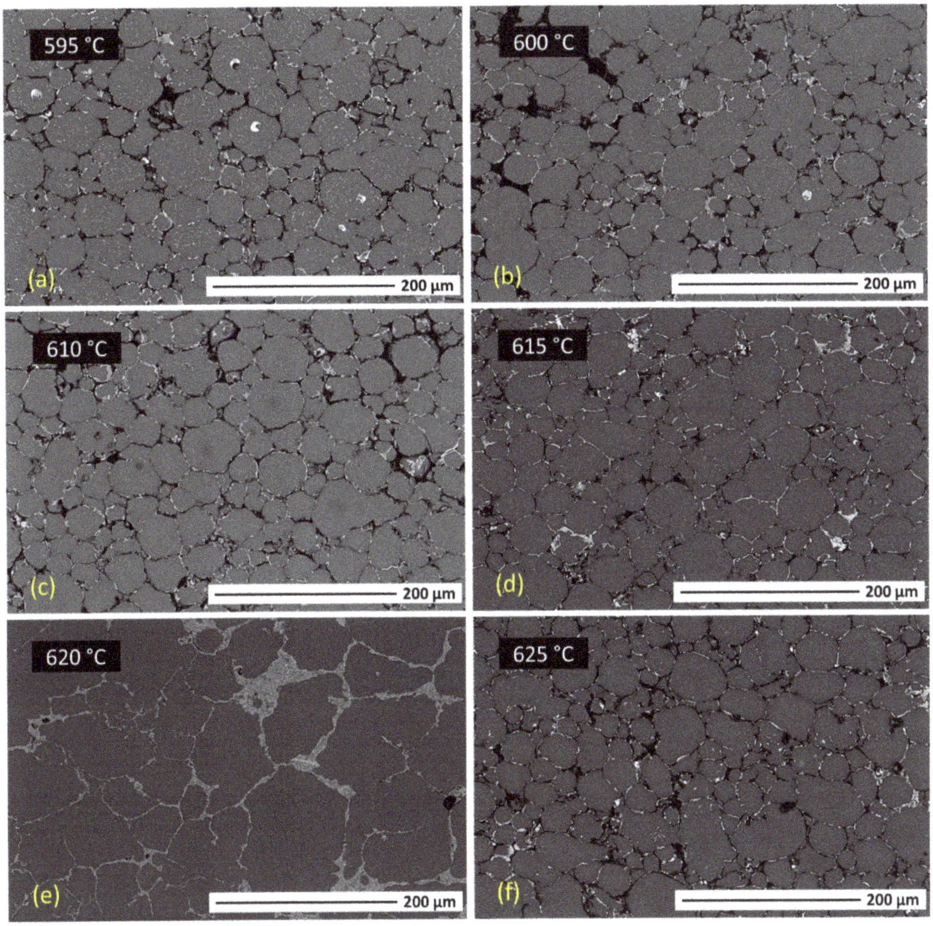

Figure 5. SEM micrographs of the cross-section of Al2024 parts sintered at different temperatures of (**a**) 595 °C, (**b**) 600 °C, (**c**) 610 °C, (**d**) 615 °C, (**e**) 620 °C, and (**f**) 625 °C.

In contrast, increasing the sintering temperature up to 625 °C again resulted in an increase in porosity. During supersolidus liquid-phase sintering, it is important to obtain an optimal volume of the liquid phase, which is conditioned by the sintering temperature and the characteristics of the powder. An excess of liquid phase results in the loss of part shape and can even lead to pore growth [31,33,34]. Therefore, compared with the sample sintered at 620 °C, the sample sintered at 625 °C suffered shape distortion and pore growth. Thus, the relative density of the sintered part at the highest temperature was reduced, which corroborates the results obtained in Figure 4.

In addition, EDS analysis of the samples sintered at different temperatures was also carried out. This semi-quantitative analysis showed that the distribution of the alloying elements (Al-94.4 wt.%, Mg-1.8 wt.%, Cu-3.8 wt.%) was homogeneous for all of them in accordance with the theoretical composition of Al2024 and the pre-alloyed powders. In particular, several analyses were performed in samples sintered at 620 °C, as seen in Figure 6. Figure 6b shows an example of one of the analyses in a selected area in the grain boundaries, which could correspond to the former liquid phase rich in copper and other alloying elements such as Mg, Mn, or Fe. Table 4 shows the results obtained in the EDS analysis seen in Figure 6b.

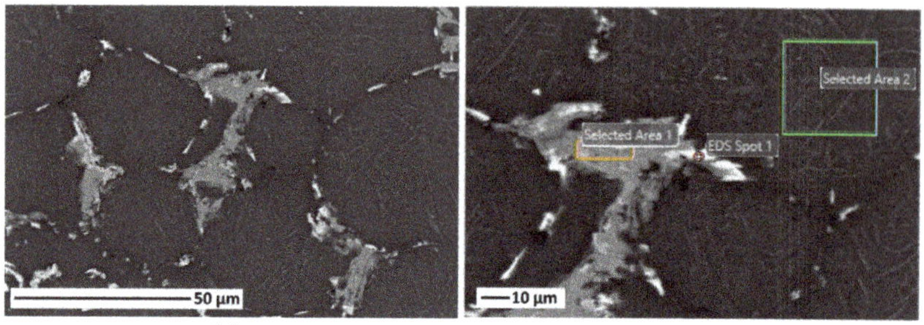

(**a**) (**b**)

Figure 6. (**a**) Micrograph of AA2024 sintered at 620 °C; (**b**) detailed micrograph of the grain boundaries.

Table 4. EDS analysis in different zones of the detailed micrograph for the Al2024 parts sintered at 620 °C.

Element	Weight %	Error %
	EDS Spot 1	
Mg	1.6	11.4
Al	60.3	7.2
Mn	1.2	15.5
Fe	1.2	16.5
Cu	35.6	3.0
	Selected Area 1	
Al	71.8	6.0
Mn	7.5	4.3
Fe	9.4	4.1
Cu	11.3	5.0
	Selected Area 2	
Mg	2.0	6.8
Al	93.9	4.0
Mn	0.6	25.6
Cu	3.5	9.0

3.2. *Solid Loading Determination and Rheological Properties of the Feedstocks*

There are numerous studies on the employment of different percentages of solid loading according to the powder properties as well as different binder systems for the manufacture of feedstocks and parts by powder injection moulding [11,18,35]. In these studies, the solid loading varies from 50 vol.% to 65 vol.%. It is important to establish the critical solid loading because during the mixing step, an excess of binder can lead to defects such as burrs or crumbling during the debinding process. On the other hand, an excess of powders could lead to defects such as porosity or high viscosity [36]. In this study, for the manufacture of the feedstocks and to be able to establish the value of the critical solid load, different solid loads from 57 to 70 vol.% were selected to study their rheological behaviour by means of the torque rheology of the feedstocks. Initially, the binder system was poured into the mixing chamber to melt for 0.166 h (10 min). After that, the amount of powder, depending on the solid content of each feedstock, was inserted into the chamber. As can be seen in Figure 7, during the mixing of all the feedstocks, there was an increase in torque when closing the mixing chamber, which progressively decreased and stabilised over time at a relatively low torque value while the mixture was homogenising. In the case of the feedstock with 65 vol.% of solid loading, the torque value was not constant after the mixing time due to the excessive amount of solid loading in the mixture. This feedstock was established as the critical solid loading. When the torque is unstable, it means that the powder loading has surpassed the critical value. For mixtures above this volume of solid loading, an inhomogeneous feedstock was created in which the binder could not accommodate more powder and did not mix well with it. It is possible to observe a final decrease in the torque value at the end of the mixing time as in the case of the feedstock with 70 vol.% of solid loading.

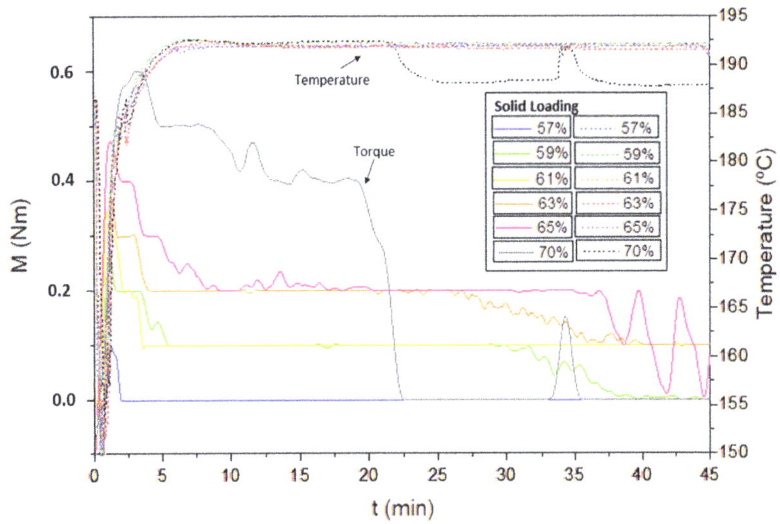

Figure 7. Variation in torque and temperature with time for PEG/CAB feedstocks with different solid loadings of Al2024 from 57 to 70 vol.%.

According to these results, feedstocks with 61% and 63 vol.% of metallic powders were established as the optimal solid loadings of those studied. These feedstocks had better rheological behaviour as they presented an optimal mixture and a better homogeneity between powder and binder for the injection process. However, the processability of the feedstocks obtained and the optimal solid loading, which is 2–5% less than the critical loading, must be taken into account [24,37,38]. Although the temperature was set at 180 °C, the average temperature reached values of around 192 °C for the production of all the feedstocks, due to the internal friction between metallic powder particles and binder during the mixing process.

In addition, the apparent viscosity of the feedstocks produced was analysed. As seen in Figure 8, all mixtures showed pseudo-plastic behaviour, whereby with increasing shear rates, the viscosity decreases. It can also be observed that by increasing the solid loading of the feedstocks, the viscosity increases slightly. PEG/CAB feedstocks were compared with a commercial feedstock of Al6061 F 19-016 (70 vol.%), showing slightly lower viscosity values due to their lower solid loadings. Although viscosities increase slightly with higher solid loadings, they always remain below the recommended viscosities for powder injection moulding (1000 Pa·s) [39]. The fluidity of the feedstocks was also studied, as can be seen in Figure 9. With this test, the pseudo-plastic behaviour of the feedstocks was again observed, where the flow capacity of the feedstocks decreased as the solid load in the mixture increased. It was observed that the feedstock with 57 vol.% of solid loading flowed too fast in the first seconds of the test, confirming that the mixture still allowed more solid load admixed, while the feedstock with 65 vol.% of solid loading scarcely presented any variation in fluidity throughout the test. It is important to obtain a balanced mixture between binder and powder, in which more homogeneous regions are created and the viscosity is lowered, providing the material with the necessary fluidity for the injection step for the PIM process [40]. Therefore, considering the results obtained for the rheological properties studied with the different solid loadings, the feedstocks with 61 vol.% and 63 vol.% of solid loading were selected as the optimal solid loading.

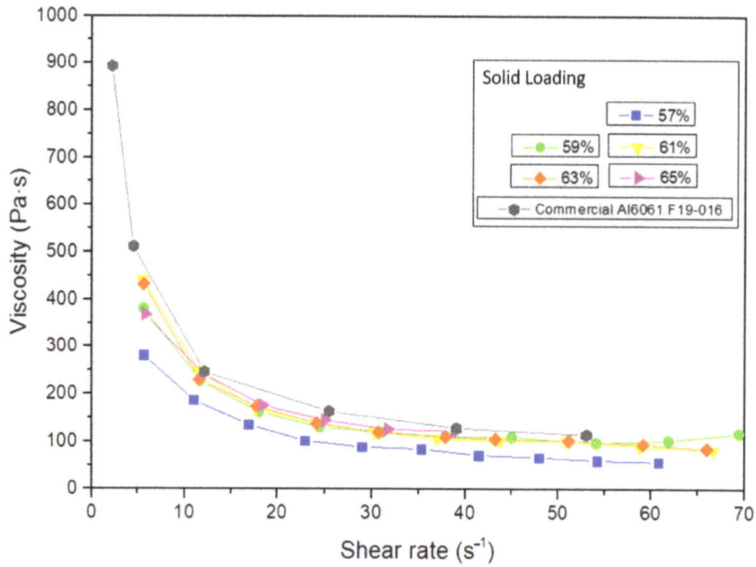

Figure 8. Apparent viscosity at different shear rates of the PEG/CAB feedstocks with a solid loading of Al2024 from 57 to 65 vol.% and commercial Al6061 F19-016.

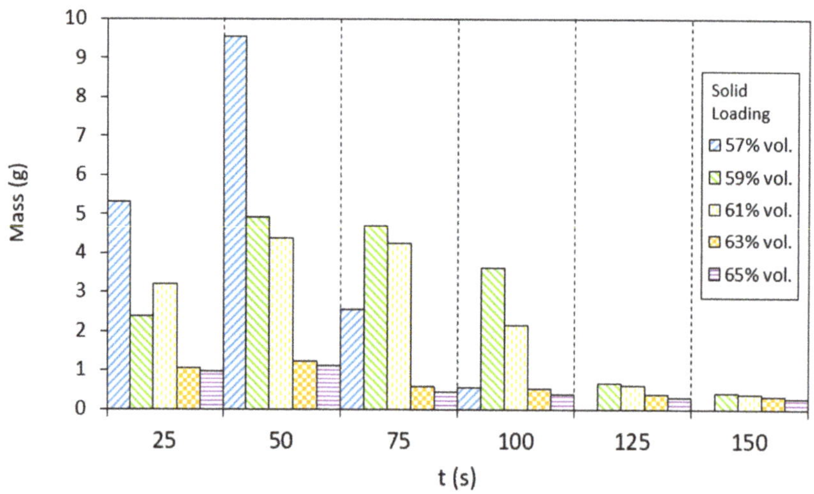

Figure 9. Variation in the fluidity of the PEG/CAB feedstocks with a solid loading of Al2024 from 57 to 65 vol.%.

3.3. Processing of the Injected Samples: Debinding and Sintering

As explained in the previous section, the optimal solid loading is 2–5% less than the critical loading, and the processability of the feedstocks obtained must be taken into account. Since it was very difficult to obtain injected parts using the feedstock with 63 vol.% of solid loading, as it had a very high solid loading, the 61 vol.% feedstock was selected as the optimal solid loading for the production of green parts and their subsequent debinding and sintering. The aluminium alloy feedstocks manufactured in the above-mentioned studies usually have a much smaller particle size (in this work, the D_{50} of our aluminium alloy powders was 45 μm) for the production of parts by the PIM process as this greatly facilitates the process. However, our work used a commercial 2024 aluminium alloy powder that was much coarser, which provided optimal solid loadings similar to those obtained in other studies with non-commercial powders designed to be finer.

Furthermore, compared to previous studies using this combination of PEG and CAB in a multicomponent binder system for obtaining feedstocks as MAX phases [38], the optimal solid loading was around 52 vol.%, whereas, in our study, the optimal solid loading was 61 vol.%. In order to know the amount of binder removed during the different debinding processes, the mass of nine samples was measured for each condition before and after debinding in water and also before and after thermal debinding. By comparing the results obtained from the masses of the samples before and after the different hours of water immersion and before and after thermal debinding in the different atmospheres, the optimum conditions of water debinding and subsequent thermal debinding were selected. Thus, taking into account previous studies [38], the two steps with the most favourable debinding conditions for this binder removal were a solvent debinding cycle in water at 60 °C with continuous agitation and an immersion time of 5 h for the removal of PEG from the green parts, followed by a thermal debinding for 1 h at 500 °C to eliminate the rest of the binder components in an argon atmosphere. An analysis of the mass variation was performed throughout the debinding cycle and, thanks to Equations (1) and (2), the amount of polymer removed was calculated. A 90 wt.% polymer removal was achieved after this two-stage debinding. For any other condition, the achieved binder removal was less than 90 wt.%. It is important to attain complete binder removal or as much as possible in the debinding stage in order to achieve high densification of the final parts in the sintering step [41]. Furthermore, an inert argon atmosphere, which reduces or prevents the formation of an Al_2O_3 layer, has been found to be more effective in eliminating this kind of polymer than any other atmosphere [35]. Nevertheless, previous studies, like those of Z. Y. Liu et al., have shown that for the sintering of aluminium alloy 6061 obtained by injection moulding, a nitrogen atmosphere is more effective [30]. That is why an argon atmosphere was not used for the sintering stage.

To verify these results from mass losses of the feedstock and confirm the removal of the polymeric binder, a thermogravimetric study of the green parts after debinding in water and after thermal debinding was carried out. Figure 10 shows the different thermogravimetric analyses for the green and brown parts for feedstock with 61 vol.% Al2024. The first analysis (black line), corresponding to the green part, shows a mass loss starting at 260 °C and ending at 405 °C; this mass loss reaches 79.9 wt.% related to the degradation of 39 vol.% of the binder in the feedstock. The second analysis (dashed blue line) belongs to the sample after PEG removal through solvent debinding. It can be seen that the part loses 12.4% less mass compared to the green sample, which corresponds to the 84.4 vol.% of the PEG removed. Finally, the third analysis (dotted brown line) shows that after the thermal cycle, all the polymeric binder has been removed from the brown parts, including the residual PEG and all the CAB from the brown part. Therefore, in accordance with the results obtained, no mass loss is expected to occur during the third analysis.

As previously seen by the study of the mass loss during the removal of the binder with Equations (1) and (2), a 90 wt.% polymer removal was obtained. However, the oxidation of the samples and the consequent weight gain during debinding may justify that a complete removal of the binder was not obtained by these equations. Thus, the thermogravimetric study of the samples was carried out in the STA. As can be seen in Figure 10, the STA test shows a complete removal of the binder, giving a more precise value of the polymer removed. These results justify the selected thermal debinding cycle of 1 h at 500 °C for the removal of CAB and the rest of the polymers from the binder, PTZ and AE, considering that the PEG polymer was previously removed by dissolution in water. In addition, after the debinding process, all the samples maintained their structural integrity and no significant defects were observed in the brown parts.

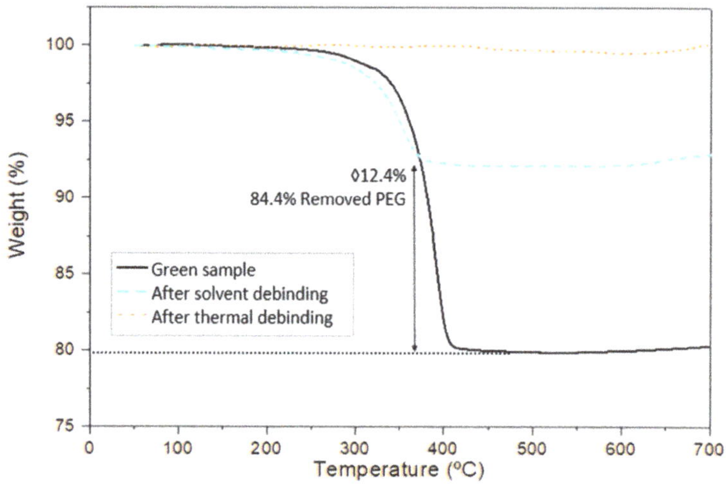

Figure 10. TGA analysis of PEG/CAB feedstock with a solid loading of 61 vol.% of Al2024 for the green simple (black line), after solvent debinding (blue dashed line), and after thermal debinding (brown dashed line).

As described above, thanks to the pressing and sintering of Al2024 powder samples and their subsequent analysis, and taking into account previous studies on the sintering of different aluminium alloys [30,32], a temperature of 620 °C was selected as the optimal sintering temperature. Therefore, the same sintering cycle was employed for the parts produced from 61 vol.% Al2024 feedstock obtained by PIM.

Figure 11 shows an example of parts after different stages of the PIM process, i.e., after injection (Green), debinding (Solvent and Thermal) and sintering (Sintered). These images show that all parts maintained their structural integrity even after the debinding and sintering steps. This indicates that both processes were carried out under optimal conditions to obtain good-quality parts without crumbling.

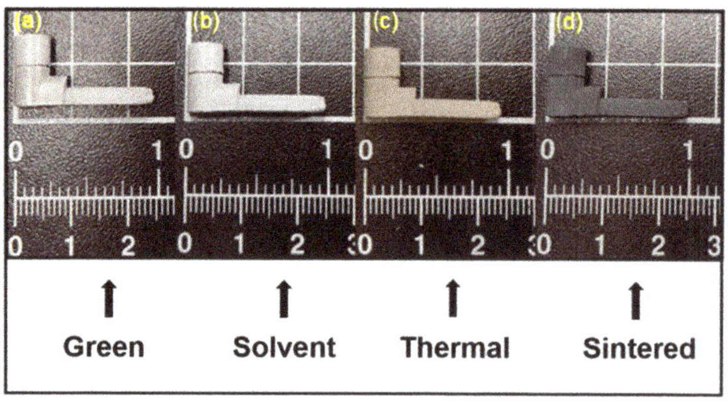

Figure 11. Injected samples of feedstock with a solid loading of 61 vol.% of Al2024 obtained at different stages: (**a**) green part, (**b**) part after solvent debinding, (**c**) part after thermal debinding, and (**d**) sintered part.

Table 5 shows the results of relative density and closed porosity obtained in green parts after the injection process, calculated with the Archimedes method. To obtain the relative density of green parts, the theoretical green density of the feedstock was calculated using the rule of mixtures. Table 5 also shows the relative density of parts after sintering. This value was obtained using the density of the Al2024 powders, measured with the helium pycnometer, as theoretical density. As can be seen, these relative density values are similar to those achieved in the pressed and sintered parts (see Figure 4), validating the PIM process for the optimisation of the debinding and sintering stages in order to obtain densified samples. The parts obtained by PIM, as well as the pressed and sintered parts, underwent shrinkage (3.11 ± 0.21% on the X axis and 1.71 ± 0.24% on the Y axis).

Table 5. Relative density and porosity of green parts and sintered parts from feedstock with a solid loading of 61 vol.% of Al2024.

Feedstock	Green Parts		After Sintering	
	ρrel (%)	**Closed Porosity (%)**	**ρrel (%)**	**Closed Porosity (%)**
Al2024 61 vol.%	79.1 ± 1.5	20.9 ± 1.5	83.7 ± 1.7	16.3 ± 1.7

This study of the optimisation of the debinding and sintering parameters of Al2024 aluminium alloy feedstocks confirms their viability for production by the PIM process. Parts of this alloy by means of 3D CEM printing technology are in production, taking into account previous studies with similar binder systems for the extrusion-based additive manufacturing of Ti_3SiC_2 and Cr_2AlC MAX phases [21].

4. Conclusions

2024 aluminium alloy parts were produced by powder injection moulding. In particular, the parameters for the production and processing of sustainable feedstocks of an Al2024 alloy with a multicomponent binder system of PEG and CAB (in a ratio of 70–30 vol.%) were determined using the PIM process. This makes it possible to comply with 2030 Agenda Sustainable Development Goal 12 on responsible and environmentally friendly production (ODS 12: Ensure sustainable consumption and production patterns). For that purpose, the optimal solid loading of the feedstocks was determined. Moreover, feedstocks produced with optimal processing conditions showed good rheological behaviour, with low torque values, low viscosity, homogeneous mixing, and pseudoplastic behaviour. The final optimal solid loading selected contained 61 vol.% of Al2024 powders. The two-step debinding process was optimised and controlled with thermogravimetric analysis for total binder removal and consisted firstly of a solvent debinding at 60 °C for 5 h and a subsequent thermal debinding at 500 °C for 1 h in an argon atmosphere. The sintering step was also optimised and determined at 620 °C for 2 h in a nitrogen atmosphere. For this purpose, SEM analysis of the cross-sections of Al2024 samples pressed and sintered at different temperatures was carried out. The sintering conditions were employed for injected samples, achieving a relative density of 87.2% with a shrinkage of 2.39 ± 0.19% on the X axis and 1.31 ± 0.31% on the Y axis. Similarly, the 61 vol.% Al2024 feedstock parts reached a relative density of 83.7% with a shrinkage of 3.11 ± 0.21% on the X axis and 1.71 ± 0.24% on the Y axis. This relative density value obtained in sintered parts can be improved by taking into account previous studies focusing on the benefit derived from the addition of small amounts of tin to the composition. Thanks to the optimisation of the feedstock carried out in this study, current work is now being developed using CEM for the production of parts of Al2024 alloy. This technology will enable cost savings compared to other laser additive manufacturing technologies or technologies using filaments.

Author Contributions: J.L.A.-G.: Conceptualisation, Methodology, Investigation, Resources, Formal analysis, Data curation, Writing—original draft. E.T.L.: Writing—review and editing. A.J.-M.: Conceptualisation, Writing—review and editing, Validation, Supervision, Funding acquisition. E.M.R.-N.:

Conceptualisation, Resources, Writing—original draft, Validation, Supervision, Funding acquisition, Writing—review and editing. All authors have read and agreed to the published version of the manuscript.

Funding: The authors would like to thank the funding provided for this research by the Regional Government of Madrid (Dir. Gral. Universidades e Investigación) through the projects PID2019-106631GB-C43 and RTC2019-007049-4 financed by the Government of Spain and the project S2018/NMT4411 (ADITIMAT-CM) financed by the Government of the Community of Madrid.

Institutional Review Board Statement: Not applicable.

Informed Consent Statement: Not applicable.

Data Availability Statement: Data are contained within the article.

Conflicts of Interest: The authors declare that they have no known competing financial interests or personal relationships that could have appeared to influence the work reported in this paper.

References

1. Mondolfo, L. *Aluminum Alloys: Structre & Properties*; Butterworth Heinemann: Oxford, UK, 1976; ISBN 978-0-408-70932-3.
2. Król, M.; Tański, T.; Snopiński, P.; Tomiczek, B. Structure and Properties of Aluminium–Magnesium Casting Alloys after Heat Treatment. *J. Therm. Anal. Calorim.* **2017**, *127*, 299–308. [CrossRef]
3. Guo, X.; Tao, L.; Zhu, S.; Zong, S. Experimental Investigation of Mechanical Properties of Aluminum Alloy at High and Low Temperatures. *J. Mater. Civ. Eng.* **2020**, *32*, 06019016. [CrossRef]
4. Graf, A. Aluminum Alloys for Lightweight Automotive Structures. In *Materials, Design and Manufacturing for Lightweight Vehicles*; Woodhead Publishing: Sawston, UK, 2020; ISBN 9780128187128.
5. Wu, M.; Liu, Y.; Wang, T.; Yu, K. Deformation Behavior and Characteristics of Sintered Porous 2024 Aluminum Alloy Compressed in a Semisolid State. *Mater. Sci. Eng. A* **2016**, *674*, 144–150. [CrossRef]
6. Zhao, N.; Ma, H.; Sun, Q.; Hu, Z.; Yan, Y.; Chen, T.; Hua, L. Microstructural Evolutions and Mechanical Properties of 6082 Aluminum Alloy Part Produced by a Solution-Forging Integrated Process. *J. Mater. Process. Technol.* **2022**, *308*, 117715. [CrossRef]
7. Christy, J.V.; Arunachalam, R.; Mourad, A.H.I.; Krishnan, P.K.; Piya, S.; Al-Maharbi, M. Processing, Properties, and Microstructure of Recycled Aluminum Alloy Composites Produced Through an Optimized Stir and Squeeze Casting Processes. *J. Manuf. Process.* **2020**, *59*, 287–301. [CrossRef]
8. Olakanmi, E.O.; Cochrane, R.F.; Dalgarno, K.W. A Review on Selective Laser Sintering/Melting (SLS/SLM) of Aluminium Alloy Powders: Processing, Microstructure, and Properties. *Prog. Mater. Sci.* **2015**, *74*, 401–477. [CrossRef]
9. Agarwala, M.; Bourell, D.; Beaman, J.; Marcus, H.; Barlow, J. Direct Selective Laser Sintering of Metals. *Rapid Prototyp. J.* **1995**, *1*, 26–36. [CrossRef]
10. Das, S.; Wohlert, M.; Beaman, J.J.; Bourell, D.L. Producing Metal Parts with Selective Laser Sintering/Hot Isostatic Pressing. *JOM* **1998**, *50*, 17–20. [CrossRef]
11. Heaney, D.F. Powders for Metal Injection Molding (MIM). In *Handbook of Metal Injection Molding*; Woodhead Publishing: Sawston, UK, 2019; ISBN 9780081021521.
12. Ghanmi, O.; Demers, V. Molding Properties of Titanium-Based Feedstock Used in Low-Pressure Powder Injection Molding. *Powder Technol.* **2021**, *379*, 515–525. [CrossRef]
13. Dhore, V.G.; Rathod, W.S.; Patil, K.N. Investigation of Mechanical Properties of Carbon Nanotubes Reinforced Aluminium Composite by Metal Injection Molding. *Mater. Today Proc.* **2018**, *5*, 20690–20698. [CrossRef]
14. Contreras, J.M.; Jiménez-Morales, A.; Torralba, J.M. Influence of the Morphology and Particle Size on the Processing of Bronze 90/10 Powders by Metal Injection Moulding (MIM). *Mater. Sci. Forum* **2007**, *534–536*, 365–368. [CrossRef]
15. Abajo, C.; Jiménez-Morales, A.; Manuel Torralba, J. New Processing Route for $ZrSiO_4$ by Powder Injection Moulding Using an Eco-Friendly Binder System. *Bol. Soc. Esp. Ceram. Vidr.* **2015**, *54*, 93–100. [CrossRef]
16. Hidalgo, J.; Abajo, C.; Jiménez-Morales, A.; Torralba, J.M. Effect of a Binder System on the Low-Pressure Powder Injection Moulding of Water-Soluble Zircon Feedstocks. *J. Eur. Ceram. Soc.* **2013**, *33*, 3185–3194. [CrossRef]
17. United Nations Sustainable Development Goals. Department of Economic and Social Affairs Sustainable Development. Available online: https://sdgs.un.org/goals (accessed on 28 November 2023).
18. Cao, P.; Hayat, M.D. Potential Feedstock Compositions for Metal Injection Molding of Reactive Metals. In *Feedstock Technology for Reactive Metal Injection Molding*; Elsevier: Amsterdam, The Netherlands, 2020.
19. Lin, D.; Sanetrnik, D.; Cho, H.; Chung, S.T.; Kwon, Y.S.; Kate, K.H.; Hausnerova, B.; Atre, S.V.; Park, S.J. Rheological and Thermal Debinding Properties of Blended Elemental Ti-6Al-4V Powder Injection Molding Feedstock. *Powder Technol.* **2017**, *311*, 357–363. [CrossRef]
20. Romanov, G.N. Liquid-Phase Sintering of Aluminum-Based Powder Alloys. *Russ. J. Non-Ferrous Met.* **2010**, *51*, 347–351. [CrossRef]

21. Tabares, E.; Kitzmantel, M.; Neubauer, E.; Jimenez-Morales, A.; Tsipas, S.A. Extrusion-Based Additive Manufacturing of Ti_3SiC_2 and Cr_2AlC MAX Phases as Candidates for High Temperature Heat Exchangers. *J. Eur. Ceram. Soc.* **2022**, *42*, 841–849. [CrossRef]
22. Singh, G.; Missiaen, J.M.; Bouvard, D.; Chaix, J.M. Copper Extrusion 3D Printing Using Metal Injection Moulding Feedstock: Analysis of Process Parameters for Green Density and Surface Roughness Optimization. *Addit. Manuf.* **2021**, *38*, 101778. [CrossRef]
23. Mukund, B.N.; Hausnerova, B. Variation in Particle Size Fraction to Optimize Metal Injection Molding of Water Atomized 17–4PH Stainless Steel Feedstocks. *Powder Technol.* **2020**, *368*, 130–136. [CrossRef]
24. Hidalgo, J.; Jiménez-Morales, A.; Torralba, J.M. Torque Rheology of Zircon Feedstocks for Powder Injection Moulding. *J. Eur. Ceram. Soc.* **2012**, *32*, 4063–4072. [CrossRef]
25. Obse, L.A.S. Determination of the Melt Mass-Flow Rate (MFR) and Melt Volume-Flow Rate (MVR) of Thermoplastics 2012. Available online: https://www.une.org/encuentra-tu-norma/busca-tu-norma/norma?c=N0049710 (accessed on 18 December 2023).
26. Olivier, D.; Aboubabky, A.; Antonia, J.M.; Manuel, T.J.; Thierry, B. Experimental and Numerical Analysis of Effects of Supercritical Carbon Dioxide Debinding on Inconel 718 MIM Components. *Powder Technol.* **2019**, *355*, 57–66. [CrossRef]
27. Schaffer, G.B.; Hall, B.J.; Bonner, S.J.; Huo, S.H.; Sercombe, T.B. The Effect of the Atmosphere and the Role of Pore Filling on the Sintering of Aluminium. *Acta Mater.* **2006**, *54*, 131–138. [CrossRef]
28. Wu, L.; Yu, Z.; Liu, C.; Ma, Y.; Huang, Y.; Wang, T.; Yang, L.; Yan, H.; Liu, W. Microstructure and Tensile Properties of Aluminum Powder Metallurgy Alloy Prepared by a Novel Low-Pressure Sintering. *J. Mater. Res. Technol.* **2021**, *14*, 1419–1429. [CrossRef]
29. Sercombe, T.B.; Schaffer, G.B. The Effect of Trace Elements on the Sintering of Al-Cu Alloys. *Acta Mater.* **1999**, *47*, 689–697. [CrossRef]
30. Liu, Z.Y.; Sercombe, T.B.; Schaffer, G.B. Metal Injection Moulding of Aluminium Alloy 6061 with Tin. *Powder Metall.* **2008**, *51*, 78–83. [CrossRef]
31. Min, K.H.; Kang, S.P.; Lee, B.H.; Lee, J.K.; Kim, Y. Do Liquid Phase Sintering of the Commercial 2xxx Series Al Blended Powder. *J. Alloys Compd.* **2006**, *419*, 290–293. [CrossRef]
32. Du, X.; Liu, R.; Xiong, X.; Liu, H. Effects of Sintering Time on the Microstructure and Properties of an Al-Cu-Mg Alloy. *J. Mater. Res. Technol.* **2020**, *9*, 9657–9666. [CrossRef]
33. Liu, J.; Lal, A.; German, R.M. Densification and Shape Retention in Supersolidus Liquid Phase Sintering. *Acta Mater.* **1999**, *47*, 4615–4626. [CrossRef]
34. German, R.M. Supersolidus Liquid-Phase Sintering of Prealloyed Powders. *Metall. Mater. Trans. A Phys. Metall. Mater. Sci.* **1997**, *28*, 1553–1567. [CrossRef]
35. Li, T. Powder Injection Molding of Metallic Parts and Structures. In *Encyclopedia of Materials: Metals and Alloys*; Elsevier: Amsterdam, The Netherlands, 2021; ISBN 9780128197264.
36. Md Ani, S.; Muchtar, A.; Muhamad, N.; Ghani, J.A. Fabrication of Zirconia-Toughened Alumina Parts by Powder Injection Molding Process: Optimized Processing Parameters. *Ceram. Int.* **2014**, *40*, 273–280. [CrossRef]
37. Barreiros, F.M.; Vieira, M.T. PIM of Non-Conventional Particles. *Ceram. Int.* **2006**, *32*, 297–302. [CrossRef]
38. Tabares, E.; Cifuentes, S.C.; Jiménez-Morales, A.; Tsipas, S.A. Injection Moulding of Porous MAX Phase Ti_3SiC_2 without Using Space-Holder. *Powder Technol.* **2021**, *380*, 96–105. [CrossRef]
39. Atre, S.V.; Weaver, T.J.; German, R.M. Injection Molding of Metals and Ceramics; In Proceedings of the SAE Technical Papers; SAE International, Princeton, NJ, USA, 1998. [CrossRef]
40. Hidalgo, J.; Jiménez-Morales, A.; Barriere, T.; Gelin, J.C.; Torralba, J.M. Capillary Rheology Studies of INVAR 36 Feedstocks for Powder Injection Moulding. *Powder Technol.* **2015**, *273*, 1–7. [CrossRef]
41. Hidalgo, J.; Jiménez-Morales, A.; Torralba, J.M. Thermal Stability and Degradation Kinetics of Feedstocks for Powder Injection Moulding-A New Way to Determine Optimal Solid Loading? *Polym. Degrad. Stab.* **2013**, *98*, 1188–1195. [CrossRef]

materials

Article

Effect of Bi on the Performance of Al-Ga-In Sacrificial Anodes

Xin Liu [1], Yufeng Lin [1], Yu Li [1,*] and Nian Liu [2,*]

1 Department of Basic, Naval University of Engineering, Wuhan 430033, China; liuxin2008dragon@126.com (X.L.); 18163558261@163.com (Y.L.)
2 Navy 91844 Troops, Guangzhou 510000, China
* Correspondence: 18162625262@163.com (Y.L.); 18162626365@163.com (N.L.); Tel.: +86-18162625262 (Y.L.); +86-18162626365 (N.L.)

Abstract: Cathodic protection is widely used for metal corrosion protection. To improve their performance, it is necessary and urgent to study the influence of metal oxides on the microstructure and performance of aluminum alloy sacrificial anodes. Taking an Al-Ga-In sacrificial anode as the research object, the dissolution morphology and current efficiency characteristics were studied by means of electrochemical testing and microstructural observation, and the influence of varying Pb and Bi contents on the performance of an aluminum alloy sacrificial anode was investigated. The test results reveal that: (1) The Al-Ga-In sacrificial anode with 4% Pb and 1% Bi contents exhibits the best sacrificial anode performance. (2) The inclusion of an appropriate Bi element content shifts the open-circuit potential in a negative direction and promotes activation dissolution. Conversely, excessive Bi content leads to uneven dissolution, resulting in the shedding of anode grains and greatly reducing the current efficiency. (3) During the activation dissolution of the aluminum alloy, the second phase preferentially dissolves, and the activation point destroys the oxide film, resulting in the dissolution of the exposed aluminum matrix. Consequently, the concentration of dissolved metal ions is reduced and deposited back on the surface of the anode sample, promoting the continuous dissolution of the anode.

Keywords: sacrificial anode; aluminum alloy; electrochemical properties; microstructure SVET

check for updates

Citation: Liu, X.; Lin, Y.; Li, Y.; Liu, N. Effect of Bi on the Performance of Al-Ga-In Sacrificial Anodes. *Materials* **2024**, *17*, 811. https://doi.org/10.3390/ma17040811

Academic Editor: Frank Czerwinski

Received: 3 August 2023
Revised: 30 January 2024
Accepted: 2 February 2024
Published: 8 February 2024

1. Introduction

In marine engineering, the corrosion of metal materials by seawater greatly reduces the utilization efficiency, service life, and safety coefficient of the materials [1–3]. Various anti-corrosion measures have been set in place, among which the sacrificial anode cathodic protection method is one of the most economical and effective anti-corrosion methods [4,5]. Sacrificial anode materials mainly include magnesium, zinc, and aluminum alloys. Aluminum alloys have emerged as the preferred sacrificial anode materials, gradually superseding traditional zinc anodes, owing to their abundant resources, cost-effectiveness, substantial theoretical capacitance, and rapid advancements in marine applications.

Research has demonstrated that adding Pb, Bi, In, Ga, and other alloy elements into aluminum alloys causes a negative potential shift, improving the efficiency of the sacrificial anodes [6–8]. Particularly, the Pb and Bi alloy elements significantly impact the aluminum anodes. For example, Li Xiaoxiang et al. have shown that Pb exhibits a notable hydrogen desorption overvoltage [9]. The enrichment and physical activation on the aluminum surface, coupled with the melting and shedding of low-melting-point compounds within the alloy, contribute to a reduction in electrochemical and resistance polarization. This process enhances the electrochemical activity of the material; hence, the aluminum alloy anode material achieves a highly negative stable potential and exhibits a low self-corrosion rate. Guo Jianzhang et al. have demonstrated that adding an appropriate amount of Bi effectively destroys the oxide film on the aluminum alloy anode surface, improves the anode activation performance, reduces the effect of grain boundary corrosion, and improves the dissolution morphology of the anode [10]. Consequently, the strengthening effect of Pb and Bi on the aluminum alloy sacrificial anodes and the influence of the microstructure on aluminum alloy anode performance have become research hotspots. Several studies have shown that Pb and Bi have synergistic effects, improving the active dissolution and current efficiency of aluminum alloys. Moreover, studies have demonstrated that the total content of both elements should not exceed 5% [11]. However, no in-depth study on Al-Ga-In alloy sacrificial anodes has been conducted yet. Therefore, the present study takes Al-Ga-In alloy sacrificial anodes as the research object. A combination of electrochemical tests and microstructure observations is employed to analyze the performance of these anodes. Moreover, this study investigated the dissolution morphology and current efficiency characteristics by determining the impact of Pb and Bi content on the performance of aluminum alloy sacrificial anodes. The findings aim to offer valuable data support for practical applications.

2. Experimental Methods

As per the literature [12], optimal active dissolution and current efficiency are observed when the total content of both elements, namely Pb and Bi, is approximately 5%. Ga and In can improve alloy dissolution activity and enhance current efficiency, and their contents should not be too high. Their reasonable content is 0.01% and 0.025%, respectively. Therefore, in this experiment, a certain amount of Ga and In were added to pure aluminum to obtain an Al-Ga-In aluminum alloy. On the basis of the Al-Ga-In aluminum alloy, the total content of Pb and Bi was set to 5%. The influence of the synergistic effect of the two elements on the sacrificial anode material was investigated by adjusting the elemental contents of Pb and Bi. Table 1 illustrates the sample labels and composition formulas.

Table 1. Anode composition formulations.

Serial Number	Pb (wt%)	Bi (wt%)	Ga (wt%)	In (wt%)	Al (wt%)
A1	5	0	0.01	0.025	Balance
A2	4	1	0.01	0.025	Balance
A3	3	2	0.01	0.025	Balance
A4	2	3	0.01	0.025	Balance
A5	1	4	0.01	0.025	Balance
A6	0	5	0.01	0.025	Balance

The aluminum alloys discussed in the paper were all prepared in the laboratory. The melting equipment was a vacuum induction suspension melting furnace, produced by Shenzhen Saimaite New Materials Co., Ltd. in Shenzhen, China, with the model number XF-1. Firstly, we placed the pure aluminum ingot in a graphite crucible and heated it in a vacuum induction suspension melting furnace. After the aluminum ingot was completely melted, we added our pre-weighed alloy elements (such as Pb, Bi, Ga, and In). Subsequently, we controlled the induction current to suspend the metal melt so that all elements were fully mixed. We poured the metal melt into the mold at an appropriate speed to form a round rod of a certain size and performed subsequent processing according to different test requirements. The use of vacuum induction suspension melting eliminates the influence of environmental media and ensures uniform mixing, resulting in reliable aluminum alloy composition.

2.1. Electrochemical Performance Test

The samples and experimental devices were prepared according to the GB/T17848-1999 [13] test methods for the electrochemical performance of sacrificial anodes. The auxiliary cathode consisted of cylindrical carbon steel; the reference electrode was a saturated calomel electrode; the working electrode contained an internal and external surface; and the working electrode area was 840 cm^2. The welded area between the auxiliary cathode and the copper rod was sealed with wax. The open-circuit potential of the anode was measured within 1 h after immersing a sample of the aluminum anode in a sodium chloride solution. A sliding rheostat was adjusted to maintain a constant current of 1 mA/cm^2. The working potential of the anode was measured every 24 h for 10 days. Then, the circuit was disconnected, the samples and the copper coulometer were removed, the corrosion pattern was recorded, and the weight changes of the anode samples and cathode copper sheets were calculated. Moreover, the anode current efficiency was calculated.

2.2. Electrochemical Behavior Test Methods

A CORRTEST CS310 electrochemical workstation was used to assess the corrosion resistance of the sample, and supporting test software was used to measure the linear polarization curve and electrochemical impedance spectrum. Epoxy resin was used to seal the edges of the sample, leaving a working surface with an area of 1 cm^2, and the sample was fixed in a test container. A three-electrode system was used in the experiment; the working electrode corresponded to the sample to be tested, the auxiliary electrode was a Shanghai Thundermagnetic 213 platinum electrode, and the reference electrode was a Shanghai Thundermagnetic 232 saturated calomel electrode. A 3.5% NaCl solution was used as the experimental electrolyte. The experimental device was kept in a terrestrial atmospheric environment, and the experimental temperature was set to 25 °C. Before starting the experiment, the sample was polished to 1000 grit, cleaned with absolute alcohol, and assembled. Once the open-circuit potential stabilized, the Tafel polarization curve and electrochemical impedance spectrum were measured. The scanning range of the Tafel polarization curve was −0.1~0.1 V (relative to the open-circuit potential); the scanning rate was 0.5 mV/s; and the frequency range of the electrochemical impedance spectrum (EIS) was 0.01~100 KHz.

2.3. Microstructure Observation and Composition Analysis

The prepared sample was cleaned with anhydrous ethanol and manually ground with sandpaper until no obvious abrasion marks could be observed. Then, electrolytic polishing corrosion was performed with a 30% KOH solution. The electrolytic voltage was set to 0.2 V, whereas the polishing time was set to 2 min. After that, the prepared samples were cleaned and dried. Finally, the microstructure was examined using a VDX 5000 3D which was made by KEYENCE in Tokyo Japan and a Leica DM 2500M metallographic microscope which was made by LEICA in Heerbrugg Switzerland. Before and after the electrochemical experiment, the sample surface was cleaned and dried with anhydrous ethanol and then put into an AURIGA scanning electron microscope to observe its microscopic morphology. Moreover, an energy-dispersive spectrometer (EDS) equipped with a scanning electron microscope (SEM) was used for element energy spectrum analysis.

2.4. Scanning Vibrating Electrode Test (SVET)

A Princeton microarea electrochemical testing system, VersaSCAN, was used to measure the potential gradient caused by the local current on the surface of the aluminum alloy sacrificial anode, as well as to describe the surface activation dissolution process of the aluminum alloy anode samples and to explore the corrosion dissolution mechanism of the Pb and Bi elements on the sacrificial anode.

3. Results and Discussion

3.1. Microstructural Observation and Compositional Analysis

The microstructure of the specimen was examined using an optical microscope, and its structural arrangement is shown in Figure 1. The microstructure of the alloy mainly consists of a gray matrix and a black granular or reticular structure. The gray matrix represents the substrate, the black particles are the second phase of segregation, and the black reticular structure is the grain boundary after erosion.

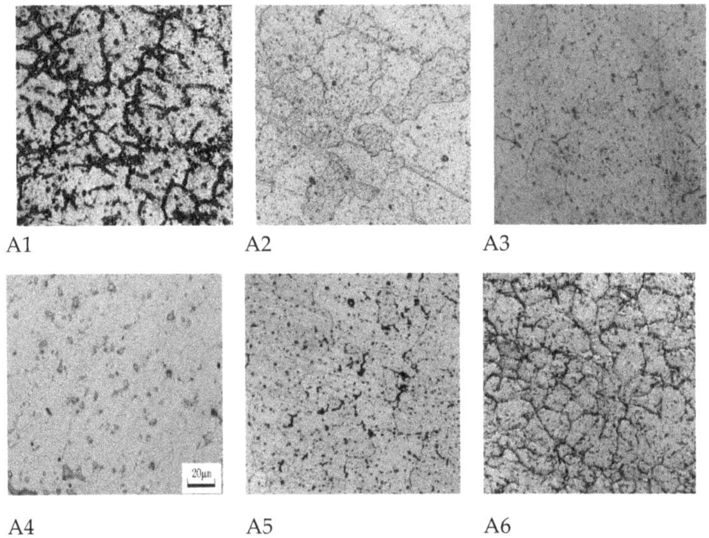

Figure 1. Metallographic structural arrangements of the specimen.

When only the Pb element is present, sample A1 exhibits a higher concentration of black particles as well as a black network structure, displaying an uneven distribution of microstructure and significant intergranular corrosion. When the amount of Bi is added, the microstructural distribution of samples A2, A3, and A4 is relatively uniform, whereas the degree of intergranular corrosion is significantly weakened. However, when the Bi content is too high, sample A6 suffers from severe grain boundary segregation and significant intergranular corrosion. This illustrates that adding an amount of Bi ranging from 1% to 3% can effectively reduce the effect of grain boundary corrosion and promote the uniformity of anodic dissolution. In samples A2, A3, and A4, the inclusion of Bi facilitates a more uniform distribution of the structure. This can mitigate the pronounced self-correction resulting from the initial segregation phase, thereby ensuring current efficiency.

3.2. Electron Probe Scanning Analysis

For samples A2, A3, and A4, electron probe technology was used to further analyze the distribution of Bi elements with different contents. Figures 2–4 represent the electron probe scanning pictures of the distribution of the Al and Bi elements in samples A2, A3, and A4.

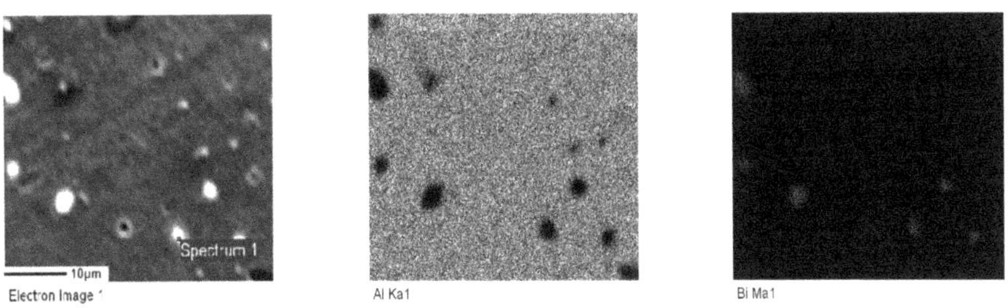

Figure 2. Element surface scanning diagram of aluminum anode sample A2.

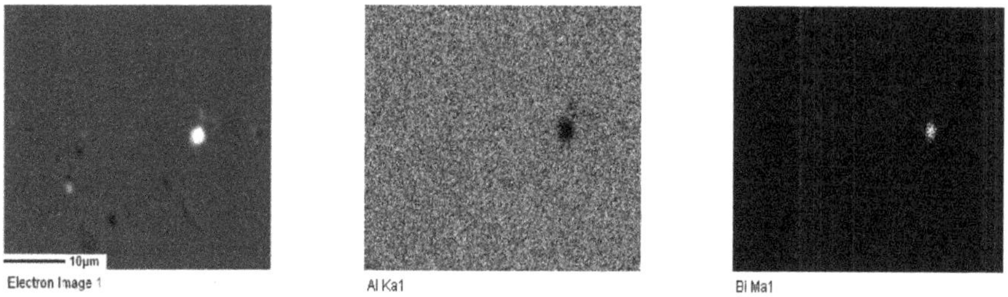

Figure 3. Element surface scanning diagram of aluminum anode sample A3.

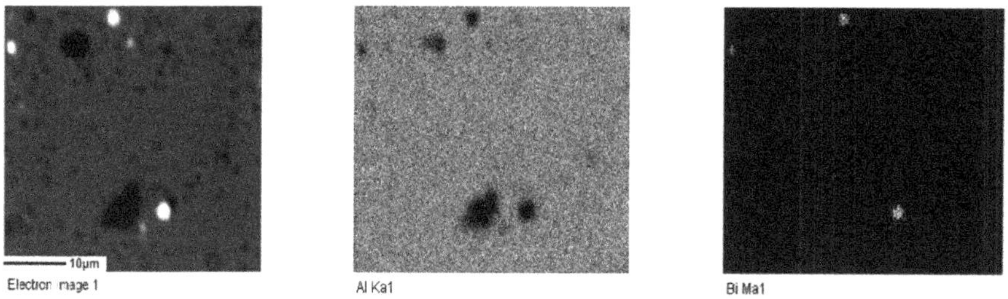

Figure 4. Element surface scanning diagram of aluminum anode sample A4.

A more uniform distribution of Bi in the alloy contributes to better structural uniformity and enhanced dissolution performance and stability during the anodic dissolution process. The electron probe scanning image reveals bright white spots, primarily corresponding to the aggregation areas of the Bi element. These spots serve as activation points in the anodic dissolution process. The bright white spots in sample A2 have a wider area and are more evenly distributed. Therefore, A2 exhibits a relatively higher number of activation points, facilitating easier activation and dissolution. It is speculated that its dissolution is more stable, leading to higher current efficiency.

3.3. *Work Performance Analysis*

Operational performance primarily encompasses the working potential of the sacrificial anode, the actual electric capacity, and the current efficiency. The working performance parameters are shown in Table 2.

Table 2. Working performance test results of sacrificial anode samples A1 to A6.

No.	Open-Circuit Potential (V)	Working Potential (V)	Actual Capacitance ($A\cdot h\cdot kg^{-1}$)	Current Efficiency (%)
A1	−0.947	−0.987~−1.034	2241.59	78.82
A2	−1.063	−0.985~−1.027	2229.56	78.36
A3	−1.356	−0.75~−1.203	1050.72	36.91
A4	−1.366	−1.216~−1.313	835.54	29.34
A5	−1.438	−1.361~−1.493	1136.57	39.89
A6	−0.998	−0.977~−1.043	1173.19	41.16

The results in Table 2 demonstrate that A3, A4, and A5 have more negative potentials in the open-circuit potential, consistent with previous microstructure analysis results. Due to the presence of more active spots in the microstructure, intergranular corrosion is likely to occur, resulting in a more negative potential. However, although A3, A4, and A5 exhibit relatively negative working positions, the working potential displays significant fluctuations. This indicates an unstable electrode surface and uneven anode dissolution, which may lead to material detachment and reduced current efficiency. The data on current efficiency further suggests that all three samples exhibit a very low current efficiency, falling below 40%. The corrosion morphology of the samples after a 30-day electrochemical performance test is illustrated in Figure 5, and the dissolution morphology is detailed in Table 3. The results demonstrate that the surface corrosion of samples A3, A4, and A5 is uneven, with material detachment occurring during the experiment. This phenomenon indicates a reduction in the current efficiency of the sacrificial anode. Moreover, it suggests that the Pb and Bi content ratio is inappropriate at this stage, leading to excessively rapid dissolution.

Figure 5. Anode surface morphology after a 240-h constant current test.

Table 3. Anode performance test results of sacrificial anode samples A1 to A6.

No.	Dissolved Morphology
A1	Pitted, pitting is more uniform, and the product is easy to fall off.
A2	Small pits, corrosion is more uniform, and the product is easy to fall off.
A3	Pockmarks, blisters, cracks, uneven erosion, and a small amount of product adherence.
A4	Blisters, cracks, grain detachment, uneven corrosion, and a small amount of product adherence.
A5	There are pits, islands, more uneven corrosion, and more product adhesion.
A6	Total corrosion, uniform corrosion, and a large amount of product adherence.

The sacrificial samples A1 and A6, with only Pb or Bi added based on their heterogeneity, had relatively positive working potentials, below −1.0 V. This is likely because Pb, being a high hydrogen overpotential element, facilitated a higher overpotential of the cathodic phase hydrogen precipitation reaction. Simultaneously, the hydrogen depolarization reaction of the cathodic phase impurities was suppressed, hindering the dissolution process of the aluminum anode. This reduction in the self-corrosion rate improves the current efficiency of the aluminum anode. When Pb and Bi coexist, the open-circuit potential of the aluminum anode is more negative. Pb and Bi exhibit synergistic effects, and the addition of an appropriate content of Pb and Bi promotes a negative shift in the open-circuit potential of the aluminum anode alloy. The A2 sample, containing an appropriate Pb and Bi content, has a relatively negative working potential of approximately −1.1 V and a higher current efficiency of nearly 80%. Therefore, A2 facilitates an excellent performance of the sacrificial anode, exhibiting a moderate open-circuit potential, uniform activation and dissolution, and a relatively high current efficiency.

3.4. Polarization Curve Analysis

Polarization curves are commonly employed to elucidate the fundamental principles of metal corrosion, reveal the underlying mechanisms of metal corrosion, and investigate strategies for corrosion control. Figure 6 represents the anodic polarization curve of the aluminum alloy specimen, whereas Table 4 contains the Tafel fitting data for the anodic polarization curve.

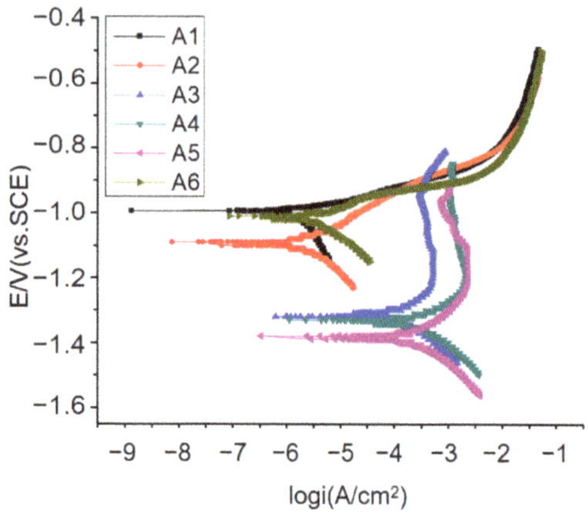

Figure 6. Polarization curves of the aluminum alloys.

Table 4. Tafel parameters of samples A1 to A6.

No.	Ba (mV)	Bc (mV)	Io (A/cm^2)	Eo (Volts)	Corrosion Rate (mm/a)
A1	44.997	437.72	2.365×10^{-6}	−0.99363	0.02583
A2	76.148	101.58	9.6624×10^{-7}	−1.0931	0.010553
A3	542.12	231.54	0.00037519	−1.3181	4.0978
A4	232.11	231.11	0.00066046	−1.3321	7.2135
A5	204.31	180.79	0.00042815	−1.3849	4.6762
A6	44.59	100.48	2.0996×10^{-6}	−1.0136	0.022932

Figure 6 shows that the anodic polarization curves of samples A1, A2, and A6 are similar and relatively smooth. Moreover, there is no obvious passivation phenomenon, indicating that the three samples have good activation performance in a 3% NaCl solution. Importantly, the three polarization curves present two obvious discharge peaks, corresponding to two discharge processes. According to the Tafel fitting results in Table 4, the self-corrosion current density, as well as the corrosion rate of the three anodes, are relatively small. However, the anodic polarization curves of samples A3, A4, and A5 are similar, exhibiting current plateau areas. In these plateau areas, variations in the anode potential have very little effect on the current change, and small current changes may result in substantial potential fluctuations. The formation of current plateaus may be attributed to the relatively high resistance of the passive film on the anode surface, making it challenging to break down when the potential is negative. When the potential exceeds a certain value, the passive film can break down, and the anode current increases rapidly. The Tafel fitting results presented in Table 4 indicate that the self-corrosion potential of samples A3, A4, and A5 is negative, while the self-corrosion current density is significantly larger, being two orders of magnitude higher than that of A1, A2, and A6. A significant number of hydrogen evolution reactions occurred during the test, generating a large number of bubbles on the anode surface. This led to grain shedding in the anode, ultimately reducing the current efficiency. Therefore, considering the corrosion rate, A3, A4, and A5 are unsuitable for sacrificial anode materials.

Among samples A1, A2, and A6, sample A2 exhibited the lowest corrosion rate and the most negative self-corrosion potential. Conversely, while sample A1 actively dissolves and serves as a sacrificial anode to protect the cathode, its lower corrosion rate indicates more even corrosion, reducing anode material shedding and improving current efficiency. This aligns with the observed trend in the change in current efficiency in the working performance of sacrificial anodes analyzed previously.

Based on the compositional analysis, the samples range from A1 to A5, reflecting a gradual decrease in the Pb element and a gradual increase in the Bi element. As the Bi element gradually increased, the potential shifted negatively, and the corrosion current density initially decreased and then increased. In general, the dissolution morphology worsened, leading to a reduction in current efficiency. Therefore, when Pb and Bi coexist, the optimal addition amount of Bi is 1.0%, indicating that A2 exhibits the best sacrificial anode performance.

3.5. EIS Analysis

The EIS of the aluminum alloy anode was used to perturb the system with small-amplitude electrical signals to obtain more kinetic and electrode interface structure information. Figure 7 shows the EIS of the aluminum alloy anode. Due to the smaller EIS radii of A3, A4, and A5, they cannot be fully displayed in Figure 7a, which is the area shown by the green circle in the figure. Enlarge the display of this area in Figure 7b. Therefore, Figure 7a shows the electrochemical impedance spectra of A1, A2, and A6, while Figure 7b shows the electrochemical impedance spectra of A3, A4, and A5.

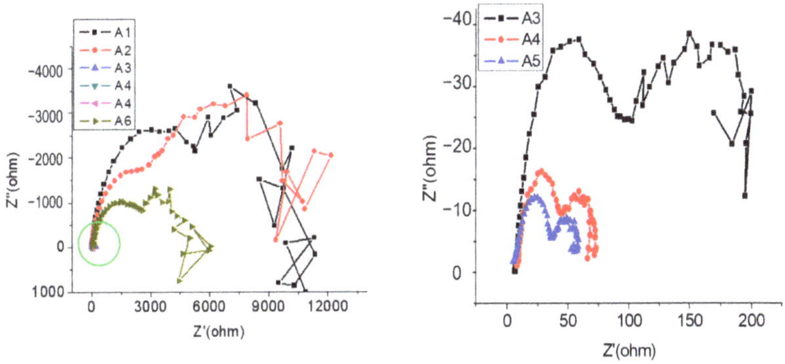

(**a**) The EIS of A1, A2, A6 (**b**) The EIS of A3, A4, A5

Figure 7. The EIS of the specimens.

The parameters of the electrochemical impedance spectra were fitted according to an equivalent circuit diagram (Figure 8) and are presented in Table 5.

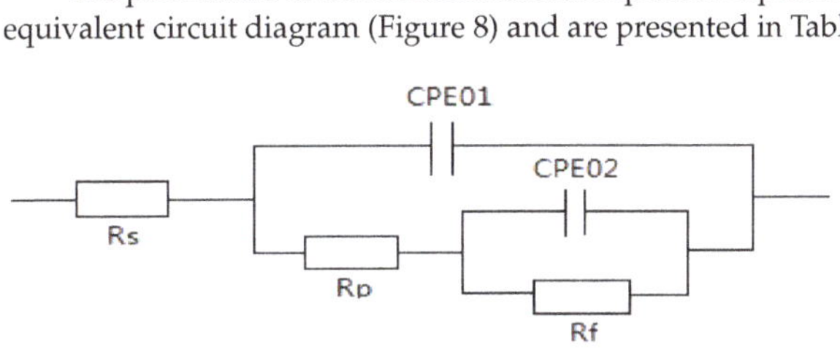

Figure 8. Equivalent circuit fitted to the EIS of the aluminum anode.

Table 5. Data table of the fitted EIS.

No.	R_S (ohm·cm^2)	Y_{01} (S·sec^n/cm^2)	n_{01} (0 < n < 1)	R_P (ohm·cm^2)	Y_{02} (S·sec^n/cm^2)	n_{02} (0 < n < 1)	R_f (ohm·cm^2)
A1	4.06	1.1×10^{-5}	0.825	6210	0.00015	1	4328
A2	2.72	1.8×10^{-5}	0.984	2848	0.000144	0.31	7985
A3	6.29	6.5×10^{-5}	0.836	81.77	0.0052	0.26	105
A4	8.26	0.01087	0.8	36.54	0.00529	0.8	23.28
A5	4.68	0.0278	0.689	34.73	0.0146	0.8	17.87
A6	9.59	2.2×10^{-5}	0.801	2450	0.0411	0.79	2187

The EIS of each sample comprises two capacitive arcs. The fitting data indicates that, with a decrease in Pb content and an increase in Bi content, the electric double-layer capacitance between the metal/passive film and the solution initially increases and then decreases. Simultaneously, the polarization resistance first decreases and then increases.

By analyzing different alloy compositions, it can be found that: (1) A6 does not contain Pb, only 5% Bi. A5 contains 1% Pb and 4% Bi, while A2, A3, A4, and A5 contain a certain amount of both Pb and Bi. (2) Compared to the samples containing Bi or Pb alone (A1 and A6), the simultaneous addition of Bi and Pb (samples A2, A3, A4, and A5) may increase or decrease the corrosion rate of the alloy, depending on the synergistic effect of Bi and Pb. (3) By comparing the corrosion rates of A2, A3, A4, and A5, it was found that when the Bi and Pb contents were 4% and 1%, respectively (sample A2), the corrosion rate was the lowest and the impedance spectrum radius was the largest.

Alloy A2 (Al-4Pb-1Bi) exhibits the largest electric double-layer capacitance and the smallest polarization resistance, representing the lowest impedance value. This suggests that A2 is the most susceptible to corrosion and dissolution, aligning with the findings of the working performance analysis and polarization curve test. The results suggest that the polarization resistance of A2 is moderate, facilitating a corrosion dissolution reaction that is not excessively corrosive. This ensures that serious self-corrosion does not occur during the continuous activation and dissolution processes, ultimately improving current efficiency. Consequently, the performance of sample A2 is superior, representing the compositional design with the best overall electrochemical performance for aluminum alloys.

3.6. SVET Analysis

The above analysis suggests that alloy A2 (Al-4Pb-1Bi) has excellent sacrificial anode performance. To further investigate this observation, a scanning vibrating electrode was employed to examine the pitting corrosion of the double electric layer of alloy A2. The focus was placed on examining the micro-area electrochemical performance during the destruction and repair of the passive film with the aim of analyzing the corrosion behavior impacting the entire material and exploring the corrosion process and mechanism of pitting corrosion and passive film destruction. Figures 9–11 show three-dimensional and two-dimensional plane views of the SVET test of sample A2 after soaking it in a sodium chloride solution (3.5% mass fraction) for 1.5 h, 3.5 h, and 5.5 h, respectively.

Figure 9 shows that after the sample was soaked for 1.5 h, an area with a significant potential difference appeared in the middle of the sample. The size of this area corresponds to approximately 2.5 mm × 2.5 mm. The second-phase particles dissolve preferentially, leading to the rupture of the passive film and thus changing the ion current above the region. After the passive film is partially broken, the aluminum substrate is exposed and starts to activate and dissolve, acting as an anode.

Figure 10 shows that the original peaks in the fourth and fifth columns of the first row and the second and third columns of the second row of the two-dimensional image disappear with time, while new peaks appear in the first column of the second row of the grid. It can be assumed that the potential of the aluminum anode gradually becomes negative as the aluminum substrate continues to be activated and dissolved. When the potential drops to a certain point, the previously dissolved alloying element ions are reduced and deposited back onto the anode surface, forming new cathodic phases, as confirmed by the appearance of new positive peaks, as illustrated in Figure 10. These newly formed cathodic phases can serve as activation points, consistently facilitating anodic activation and dissolution. The peaks in the first row and column persist, albeit with slight upward shifts and expanded mapping regions. It is hypothesized that the rupture of the oxide film in this region is followed by the chemical dissolution of the surrounding aluminum matrix. The second-phase particles in this area are slightly larger, gradually revealing the shape of the second phase.

With increasing corrosion time, the continuous aggregation of anions and cations leads to a gradual increase in the potential difference between the second interaction and the matrix, indicating that the corrosion activation process is deepening. Figure 11 shows that after immersion for 5.5 h, the maximum potential difference in this area is more than nine times that of the first scan and more than two times that of the second scan. The blocky, warm color areas in the fourth and fifth columns of the second row indicate that a large number of metal ions are deposited back to form a cathode phase. A warm-colored area is always present in the first row and column. The area gradually expands during the scanning process, and the complete shape of the warm-colored area is not visible in the scanning area. This is because the second-phase particles in this region are larger, have more contact with the substrate, and are challenging to dissolve or separate.

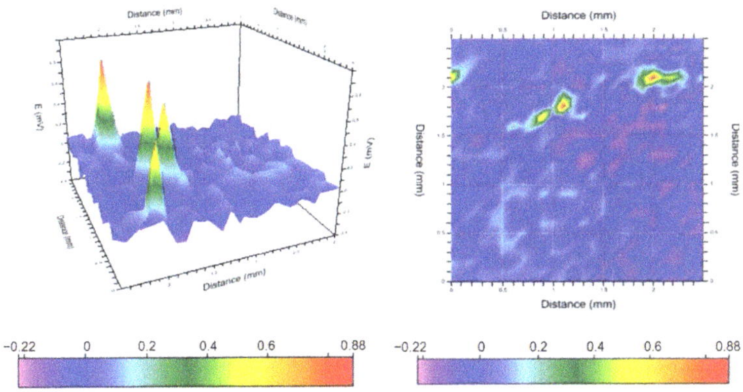

(**a**) Three-dimensional diagram (**b**) Two-dimensional diagram

Figure 9. Voltage distribution diagram after 1.5 h of immersion.

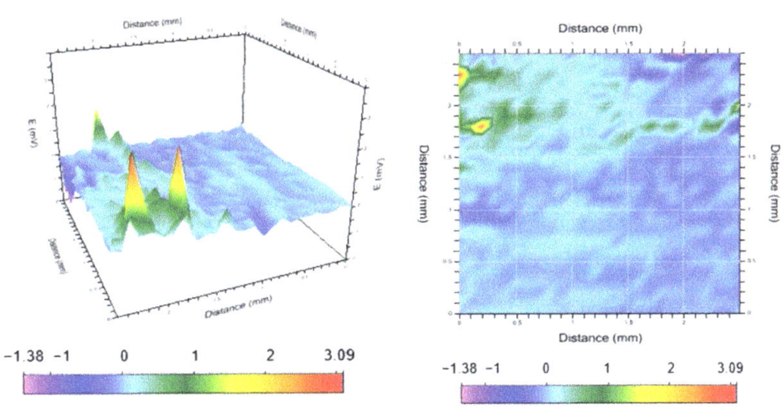

Figure 10. Voltage distribution after soaking for 4.5 h.

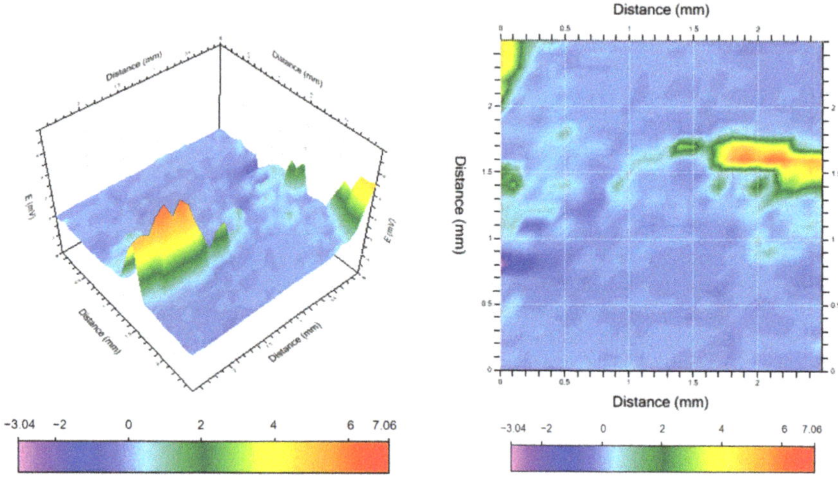

Figure 11. Voltage distribution after 5.5 h of soaking.

In summary, the activation and dissolution processes of the aluminum alloy can roughly be described as follows: (1) starting from the preferential dissolution of the second phase, as the reaction proceeds, the passivation film undergoes a local fracture, the aluminum substrate is exposed, and activation and dissolution begin; (2) as the reaction continues, the small-sized second phase is dislodged and dissolved, and the potential of the aluminum anode is shifted. After the metal ions dissolved in the medium are deposited back to the anode surface, a new cathodic phase is formed, serving as a new activation point for the continuous dissolution of the aluminum anode. (3) If there is a more negative second phase or inclusions in the alloy, severe pitting corrosion will occur.

4. Conclusions

The Al-Ga-In sacrificial anode has the best sacrificial anode performance when 4% of Pb and 1% of Bi are added.

An appropriate Bi element content can shift the open-circuit potential negatively and promote activation dissolution. Excessive Bi elements are prone to uneven dissolution, resulting in the shedding of anode grains and greatly reducing the current efficiency.

During the activated dissolution of aluminum alloys, when the second phase preferentially dissolves or the activation point destroys the oxide film, the initial dissolution exposes the aluminum substrate. Subsequently, the already dissolved metal ions are reduced and deposited back onto the surface of the anode sample, facilitating the continuous dissolution of the anode.

Author Contributions: Conceptualization, X.L.; Formal analysis, Y.L. (Yufeng Lin); Investigation, Y.L. (Yu Li); Resources, N.L. All authors have read and agreed to the published version of the manuscript.

Funding: This research was funded by the Naval Engineering University's Independent Research and Development Program (grant number 425317T005) and Key R&D Projects in Hubei Province (grant number: 2022BID014).

Institutional Review Board Statement: Not applicable.

Informed Consent Statement: Informed consent was obtained from all subjects involved in the study.

Data Availability Statement: Data are contained within the article.

Conflicts of Interest: The authors declare no conflict of interest.

References

1. Chantarojsiri, T.; Soisuwan, T.; Kongkiatkrai, P. Toward green syntheses of carboxylates: Considerations of mechanisms and reactions at the electrodes for electrocarboxylation of organohalides and alkenes. *Chin. J. Catal.* **2022**, *43*, 3046–3061. [CrossRef]
2. Rezaei, A. Artificial neural network modelling to predict the efficiency of aluminium sacrificial anode. *Corros. Eng. Sci. Technol.* **2023**, *58*, 747–754. [CrossRef]
3. Mujezinović, A.; Martinez, S. Sacrificial Anode Cathodic Protection System Modelling Using Direct Boundary Element Method. *BH Electr. Eng.* **2022**, *16*, 25–30. [CrossRef]
4. Isra, M.; Huzni, S.; Israr, B.M.I.; Fonna, S. The Effectiveness Evaluation of Sacrificial Anode in Reinforced Concrete that has been Installed for One Year. *Key Eng. Mater.* **2022**, *6639*, 187–195. [CrossRef]
5. Rees, A.B.; Gallagher, A.; Wright, L.A.; Wood, J.; Cathery, T.; Harrison, B.; Down, C.; Comber, S. Leisure craft sacrificial anodes as a source of zinc and cadmium to saline waters. *Mar. Pollut. Bull.* **2020**, *158*, 111433. [CrossRef] [PubMed]
6. Hariesh, S.P.; Dhanyanth, P.V.; Kughan, S.A.; Babu, D.K.; Vaira, V.R.; Sathiya, P. Influence of temperature and thermal cycles on the corrosion mechanism of wrought AZ91D magnesium alloy in simulated sea water solution. *Corros. Mater. Prot. J.* **2023**, *67*, 66–77.
7. Jima, S.W.; Melesse, E.Y.; Endale, A.T. Comparison of the Removal Efficiencies for Electro-Flotation and Electrocoagulation Wastewater Treatment Methods to Treat Effluents Discharged from Electroplating Industries. *Int. J. Eng. Res. Afr.* **2023**, *6785*, 67–82. [CrossRef]
8. Soltanpour, M.; Boroujeny, B.S.; Nourbakhsh, A.A.; Akbari-Kharaji, E. Microstructural and Corrosion Characteristics of Al-Zn-In Sacrificial Anode Fabricated by Cooling Slope Casting Through Response Surface Methodology: A Modeling Study. *Metallogr. Microstruct. Anal.* **2023**, *12*, 116–133. [CrossRef]
9. Ma, Z.; Li, X. The microstructure and electrochemical properties of Al-Mg-Pb-Sn-Ga alloy anode material. *J. Hunan Univ. Sci. & Technol. (Nat. Sci. Ed.)* **2011**, *26*, 73–79.
10. Guo, J.; Hu, C.; Zhang, H. Electrochemical behaviour of Al-Zn-Ga and Al-In-Ga alloys in chloride media. *Corros. Sci. Prot. Technol.* **2019**, *31*, 489–494.
11. Elsayed, A.; Nofal, A.; Attia, G. Development of metal oxide incorporated Al-Zn-Sn sacrificial anodes processed by stir casting and heat treatment. *J. Solid State Electrochem.* **2022**, *26*, 2659–2672. [CrossRef]
12. Safwat, S.M.; Mohamed, N.Y.; El-Seddik, M.M. Performance evaluation and life cycle assessment of electrocoagulation process for manganese removal from wastewater using titanium electrodes. *J. Environ. Manag.* **2023**, *328*, 116967. [CrossRef] [PubMed]
13. *GB/T 17848-1999*; Test Method for Electrochemical Properties of Sacrificial Anodes. State Bureau of Quality and Technical Supervision: Beijing, China, 1999.

materials

MDPI

Article

Influence of Aspect Ratio on the Flexural and Buckling Behavior of an Aluminium Sandwich Composite: A Numerical and Experimental Approach

Ganesh Radhakrishnan [1], Daniel Breaz [2,*], Al Haitham Mohammed Sulaiman Al Hattali [1], Al Muntaser Nasser Al Yahyai [1], Al Muntaser Nasser Omar Al Riyami [1], Al Muatasim Dawood Al Hadhrami [1] and Kadhavoor R. Karthikeyan [3,*]

1 Department of Mechanical Engineering, College of Engineering & Technology, University of Technology & Applied Sciences, Nizwa P.O. Box 477, Oman; ganesh.radhakrishnan@nct.edu.om (G.R.); s22s161230@nct.edu.om (A.H.M.S.A.H.); s72s179@nct.edu.om (A.M.N.A.Y.); 21s13124@nct.edu.om (A.M.N.O.A.R.); s22s1775@nct.edu.om (A.M.D.A.H.)
2 Department of Mathematics, "1 Decembrie 1918" University of Alba Iulia, 510009 Alba Iulia, Romania
3 Department of Applied Mathematics and Science, National University of Science & Technology, Muscat P.O. Box 620, Oman
* Correspondence: dbreaz@uab.ro (D.B.); karthikeyan@nu.edu.om (K.R.K.)

Abstract: In the field of engineering materials, lightweight and ultra-lightweight composites are used in real time to a greater extent, with high-performance targeting for tailor-made systems in aerospace, automotive, and biomedical applications. Sandwich composites are among the most popular lightweight materials used in structural and vehicle-building applications. In the present investigation, one such sandwich composite laminate composed of aluminum face sheets and a high-density polyethylene core was considered to analyze sandwich composites' flexural and buckling behavior experimentally and numerically. The influence of aspect ratios, such as length to thickness and width to thickness, on the flexural and buckling performance of sandwich composite laminates was explored in the study. Laminates with different widths, namely, 10, 12, and 15 mm, and a uniform thickness and length of 3 mm and 150 mm, respectively, were used for flexural analysis, whereas laminates with widths of 10, 12, and 15 mm and a uniform thickness and length of 3 mm and 350 mm, respectively, were used for buckling analysis. The geometrical influence of the laminates on mechanical performance was studied through performance measures such as critical bending load, flexural stiffness, inter-laminar shear stress, and critical buckling load. A significant influence of aspect ratio on the mechanical behavior of the laminates was observed using both experimental and numerical approaches. Flexural behavior was observed to be better at greater widths, namely, 15 mm, and with a minimum support span of 90 mm due to reduced spring back effects and increased bending resistance. A maximum width of 15 mm allowed for a higher buckling load capacity similar to that of bending resistance. A critical buckling load of 655.8 N seemed to be the maximum and was obtained for the highest aspect ratio, b/t = 5. The soft core and ductile metal face sheets offered combined resistance to both bending and buckling. A lower aspect ratio (span to thickness) rendered these sandwich laminates better in terms of both bending and buckling.

Keywords: sandwich composite; laminate; aspect ratio; flexural strength; buckling; inter-laminar strength

check for updates

Citation: Radhakrishnan, G.; Breaz, D.; Al Hattali, A.H.M.S.; Al Yahyai, A.M.N.; Al Riyami, A.M.N.O.; Al Hadhrami, A.M.D.; Karthikeyan, K.R. Influence of Aspect Ratio on the Flexural and Buckling Behavior of an Aluminium Sandwich Composite: A Numerical and Experimental Approach. *Materials* **2023**, *16*, 6544. https://doi.org/10.3390/ma16196544

Academic Editor: Tomasz Sadowski

Received: 7 September 2023
Revised: 24 September 2023
Accepted: 2 October 2023
Published: 3 October 2023

1. Introduction

In general, composite materials have revolutionized the materials field in terms of unique, tailor-made features and high performance. Every day, new classes of composite materials are being developed all over the world, with attractive features such as being ultra-lightweight, having improved mechanical and thermal performance, etc. One such popular category of lightweight composite materials is sandwich composites. Sandwich composite laminates are a kind of composite material wherein metal–polymer–metal sandwich sheets are bonded together with an adhesive. These composites consist of two thin, lightweight metal face sheets and a thick polymer core. The polymer core has low strength, but its high thickness results in higher flexural resistance with low density. The most popularly used core materials include polyurethane, polyethylene, polystyrene, honeycombs, etc. Thermoplastics, thermoset polymers, and sheet metals are used as skin materials. The important feature of a sandwich composite is that its exterior surfaces resist the loads caused by bending or compression, whereas the inner core material resists the load caused by shearing. Basically, sandwich composite structures are types of anisotropic materials; therefore, the strength of the material depends on the nature of the applied load. The selection of materials and their dimensions govern an object's resistance capability against loading. Sandwich composite structures are mainly used for applications where stiffness is considered a significant factor. Analyzing the mechanical behavior of sandwich composite structures with respect to the components used in the structure and the relevant loading conditions will reveal their significant influence on performance when applied to a real time environment. The properties of the materials used for the core and face sheets and their dimensions have a great influence on the performance of a composite. Simulated analysis of the mechanical behavior of these sandwich composites gives the opportunity to find an appropriate material and determine its usefulness.

In the present work, aluminum sandwich composite (ASC) laminates were selected due to their availability and high performance in real-time applications. Aluminum sandwich composite laminates are used in structural and building applications regarding doors, covering external structures, etc., to provide weather resistance and soundproofing [1–3]. They have been investigated for their performance against static flexural and buckling loading. Flexural and buckling behaviors were chosen because failures in structural and building applications are more critical due to bending and buckling. Shear between the face sheets and the core material was observed due to a higher order of bending under transverse loading and lateral buckling under a compressive load [4,5]. The simulated flexural and buckling behavior of the ASC laminate was compared with experimental work to confirm the consistency and accuracy of the results obtained. The key factor, the aspect ratio, which is the ratio between two dimensions of a work piece material, is considered in the flexural behavior study of ASC laminates. Aspect ratios have a significant influence on the flexural behavior of composite laminates. In this study, aspect ratios, i.e., the ratios of width to thickness and length to thickness, are considered and investigated to determine their influence. Ali Isiktas et al. [6] analyzed the cracks on carbon-fiber-reinforced Al laminate formed during bending and found that the dimensions of the cracks increased with the increase in the thickness of the laminate. Similar observations were made by Tang et al. [7] while assessing the bending performance of carbon fiber epoxy laminate. Cracks with slips appeared on the surface of sandwich laminate due to heterogeneity in construction. PVC foam glass-fiber-reinforced polymer laminate was subjected to transverse loading, and multiple failure modes were explored, such as fiber failure, shear failure between the matrix and the fiber, delamination, debonding at the interface, and foam failure. A few years back, Davies [8] reviewed the structural design of elements in the sandwich panel, which consist of two metal faces separated by a lightweight core. The buckling behavior of the panel was investigated in the study through classical and numerical approaches, and the results showed that a greater contribution of resistance to buckling was made by the core material compared to the metal faces. The simplification of the design aspects changed the investigation from a quite complex analysis to an analysis with appropriate results. A numerical analysis of sandwich panels with a new type of element was proposed by Ya Ou et al. [9]. The element they used comprised two face layers connected by another layer. The face layer was considered to serve as a beam, and the intermediate layer was considered to serve as a spring. Researchers and academicians working in the field of sandwich composite structures have contributed many useful outcomes, but there is still a wide scope to pursue with respect to research in this area in order to find novel material combinations, thereby making the resulting product suitable for a desired application. Most of the contributions witnessed in the area of investigation have concentrated on natural- or synthetic-fiber-reinforced tailor-made polymeric composites, and very few research outcomes regarding commercially available sandwich panels have been published. The highlights of this work include the investigation of the bending and buckling behavior of commercially available aluminum sandwich composite laminates in order to recommend them for structural applications, where high specific performance is a desired factor. The novelty of this work is that it considers a geometrical factor, the aspect ratio, as a variable for comparison with other different characterization parameters analyzed in the study. Based on a detailed literature review, it was assumed in the investigation that there was perfect bonding between the face sheets and the core. It was also assumed that there was no delamination between the layers. The metal face sheets were assumed to be elastic at all times. The research objective of this study was to investigate the influence of the aspect ratio of sandwich laminates on their flexural and buckling performance, which, in turn, would help us to arrive at optimized dimensions of the laminates. This objective was chosen to determine how to limit the wastage of a material and thereby increase its specific strength.

2. Experimentation

2.1. Materials

Commercially available aluminum sandwich composite (ASC) laminates with a 2.4 mm thick, black, high-density polyethylene core and aluminum sheet metal pieces of 0.3 mm thick placed on either side of the core as a face material were used in this study. The ASC laminate specimens prepared for three-point bending test had rectangular cross-sections with three different widths corresponding to 10, 12, and 15 mm and a fixed length and thickness of 150 mm and 3 mm, respectively. A schematic sketch of the specimen's cross section is shown in Figure 1. The dimensions of the specimen were chosen based on the sizes of different types of commercially available aluminum sandwich panels available on the market. The dimensions suitably matched the experimental set up.

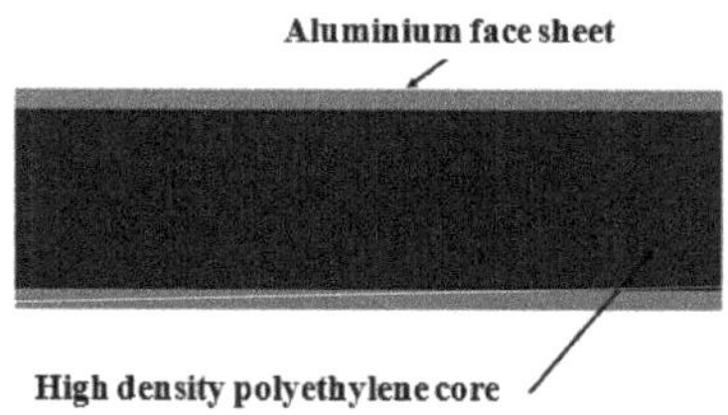

Figure 1. Schematic sketch of the cross section of sandwich composite laminate.

2.2. Methodology

2.2.1. Experimental Analysis

The flexural behavior of the laminate was tested using a universal Testing Machine (UTM) with a 20 kN capacity supplied by Gunt Hamburg, Brighton, Germany. A concentrated load was applied on the specimen manually at the mid span in slow and gradual increments. The data acquisition system of the machine displayed the deflection versus load as output, from which maximum bending load was estimated. Each experiment was repeated thrice to confirm the accuracy and consistency of the results. This procedure reduces the number of errors that can occur during experimentation. The experimental setup used in the study is shown in Figure 2a,b. The influence of aspect ratio on the flexural behavior of the sandwich laminate was investigated in the study. Aluminum sandwich composite (ASC) laminates were also subjected to buckling test under various end conditions in a buckling behavior machine with a capacity of 2500 N (supplier: Gunt Hamburg, Brighton, Germany). The specimens were prepared with three different widths, namely,

15, 12, and 10 mm, and had a common length of 350 mm. Buckling tests were carried out on the specimens to determine the end conditions based on Euler theory. The specimens were prepared accordingly for the end conditions: pinned–pinned, pinned–fixed, and fixed–fixed. Each test was repeated three times with different specimens to ensure the consistency of results obtained. The output measures such as maximum buckling load and corresponding deflection were noted.

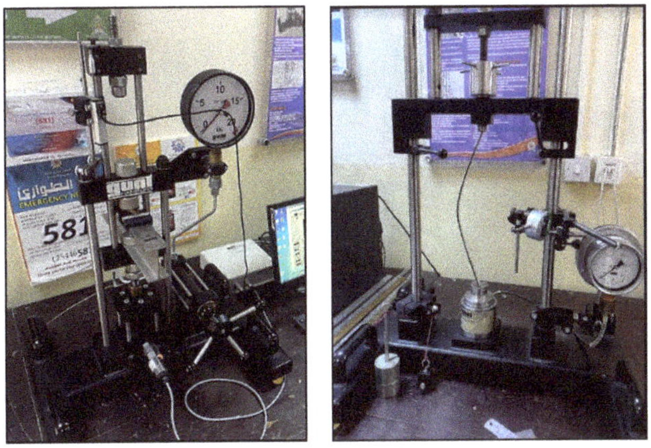

(a) (b)

Figure 2. Experimental set up for (**a**) flexural test and (**b**) buckling test.

2.2.2. Numerical Analysis

The numerical study model is shown in Figure 3. Since the specimens were very thin and lightweight, the buckling load was observed to be very minimum. In the numerical analysis, the flexural and buckling experiments were modeled using SOLIDWORKS (Version 2018, Dassault Systems, Waltham, MA, USA), with details closer to the experimentation, and imported to ANSYS (ANSYS workbench 2012, ANSYS Inc., Canonsburg, PA, USA). The aluminum sandwich composite laminate was modelled in ANSYS as a single component with three layers across the width, namely, a polyethylene core in the middle that was 2.4 mm thick and aluminum face sheets on either side that were 0.3 mm thick each. All the constituents were assumed to be cohesive in the numerical analysis. In the numerical analysis, the contact between the contacting surfaces, such as the beam, being a sandwich laminate, and the supporting and loading rollers, being made of steel, was considered frictionless. The boundary conditions and constraints were applied in the numerical analysis, matching the conditions applied during the experimentation. The cross-section was symmetrical, and the loading plane was also a symmetrical section. The loading plane varied with respect to the width of the laminate, that is, 10, 12, and 15 mm, whereas the length was maintained constant at 150 mm. The supports were simply supported, with 'Z' translation free and 'X' rotation free. Fine mesh with standard 10-node tetrahedron elements with a size of 1 mm was used in the numerical analysis for both core and face sheets in order to reduce the computational effort. Constant mesh density was used in the analysis for better resolution. The material properties of the core and face layers were used in the numerical analysis, and they were the same as those used in experimentation. The adhesive interfaces between the metal face layer and polymer core were modelled in ANSYS as cohesive contact elements with zero thickness. The team of Massimo et al. [10] made a similar assumption and followed a similar procedure in their study. Also, nonlinear analysis of the structures with implemented geometric imperfections was conducted by Pawel [11] to study the effect of compression load eccentricities on buckling behavior, whereas in this study, an axial compressive load was applied instead of eccentric load.

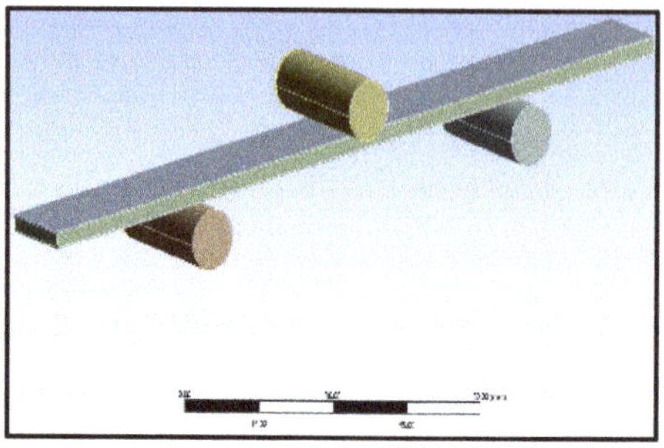

Figure 3. Flexural test model for numerical analysis.

3. Results and Discussions

The experimental and numerical values of the performance measures obtained are shown in Table 1. The average values of the performance measures obtained in all the trials are presented in the table. The bending load was maximum at around 1.5 kN for the aspect ratios L/t and b/t corresponding to 30 and 5, respectively. It can be observed in Figure 4 that there was a drastic drop in bending load with an increase in the aspect ratio, L/t, from 30 to 36.67 irrespective of the width of the laminates. On the other side, when L/t increased from 36.67 to 43.33, there was a marginal or negligible rise in the bending load. The greater the width of the laminate, the higher the resistance offered by the specimen against bending [12]. This was achieved through the combined resistance of the polyethylene core and the aluminum face sheets. The thickness of the core used in the construction of the laminate was 2.4 mm, and the thickness of the face sheets used on either side was 0.3 mm. The greater the thickness of the core, the greater the resistance offered against deformation. Figure 5 reveals that the deflection measured at the midpoint of the laminate during the flexural test increased steadily with the increase in the aspect ratio, L/t, from 30 to 43.33. An increase in the support span led to a decrease in the stability of the laminates against transverse bending load irrespective of the width of the laminates [13–16]. The lowest order of deflection was observed for the wider laminate due to the higher modulus offering higher resistance to bending. The flexural stiffness of the sandwich composite laminates was measured to study the resistance offered by the laminate during deformation and plotted against the support span; this information is shown in Figure 6. The observations regarding the flexural stiffness of the laminates confirmed that the flexural behavior was similar to that obtained through the measurement of deflection. A steady drop in flexural stiffness with an increase in support span was noticed. Since this study focused on a sandwich composite laminate, the important characteristic that needed to be considered was inter-laminar shear stress due to the adhesive contact surfaces available on both sides of the laminate [17–20]. These adhesive layers often tend to tear off, increasing the possibility of failure. The variation in inter-laminar shear stress against support span is shown in Figure 7. It was observed that there was a steady drop in inter-laminar shear stress as the support span increased from 90 to 110 mm and further increased to 130 mm; the variation in inter-laminar shear stress was very marginal and constituted a negligible quantity. The shear deformation was enhanced significantly with an increase in the support span. This may be attributed to the increased slope of the laminates under loading and the increase in support span. The constituents of the sandwich composite laminate, such as the soft polyethylene core and the metal face sheets on either side, restricted the shear deformation when the support span increased beyond 110 mm. The adhesive layer connecting the face sheet and the core on either side influenced the behavior over the entire span of the laminate [21–23].

Table 1. Numerical vs. experimental results.

S. No.	L/t	b/t	Maximum Bending Load (kN)			Maximum Deflection (mm)			Flexural Stiffness (kN/mm)			Inter-Laminar Shear Stress (kN/mm²)		
			Expt.	Num.	Error (%)	Expt.	Num.	Error (%)	Expt.	Num.	Error (%)	Expt.	Num.	Error (%)
1	30	5	1.34	1.63	17.3	10.19	9.91	2.7	0.13	0.16	−23.1	0.022	0.027	−22.7
2	30	4	1.02	1.26	18.4	9.71	9.93	−2.3	0.10	0.12	−20.0	0.021	0.026	−23.8
3	30	3.3	1.00	1.01	1.1	11.35	8.42	25.8	0.08	0.12	−50.0	0.025	0.026	−4.0
4	36.67	5	0.75	0.78	4.8	10.36	13.59	−31.2	0.07	0.05	28.6	0.012	0.013	−8.3
5	36.67	4	0.62	0.80	22.7	10.73	13.12	−22.3	0.05	0.06	−20.0	0.012	0.017	−41.7
6	36.67	3.3	0.51	0.61	15.4	10.72	12.97	−21.0	0.04	0.04	0.0	0.012	0.015	−25.0
7	43.33	5	0.75	0.84	10.8	17.24	19.67	−14.1	0.04	0.04	0.0	0.012	0.014	−16.7
8	43.33	4	0.77	0.78	1.1	22.11	19.87	10.1	0.03	0.04	−33.3	0.016	0.016	0.0
9	43.33	3.3	0.65	0.75	13.3	22.27	19.31	13.3	0.02	0.03	−50.0	0.016	0.019	−18.8

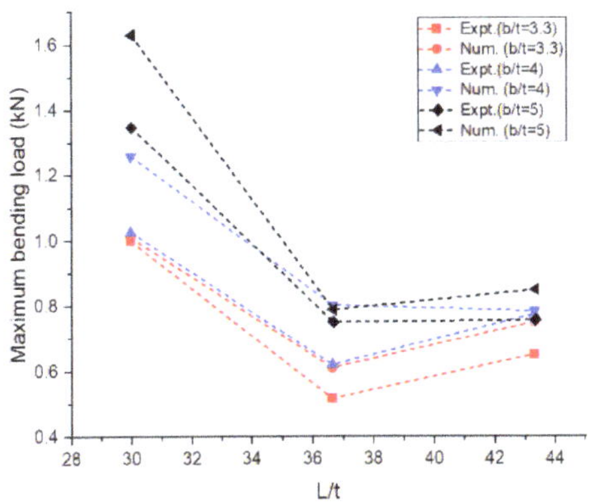

Figure 4. Aspect ratio vs. bending load: numerical and experimental.

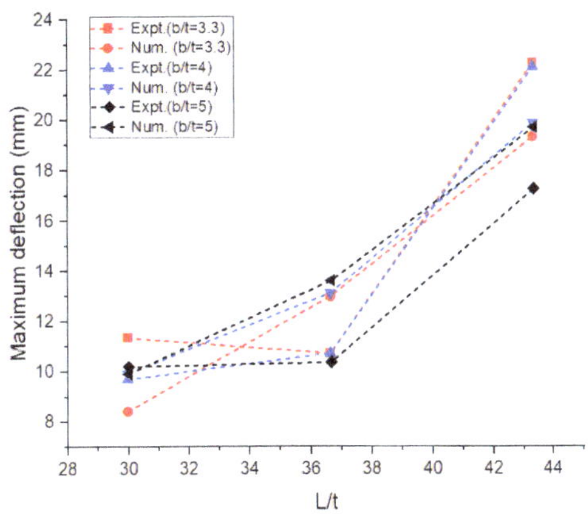

Figure 5. Aspect ratio vs. deflection: numerical and experimental.

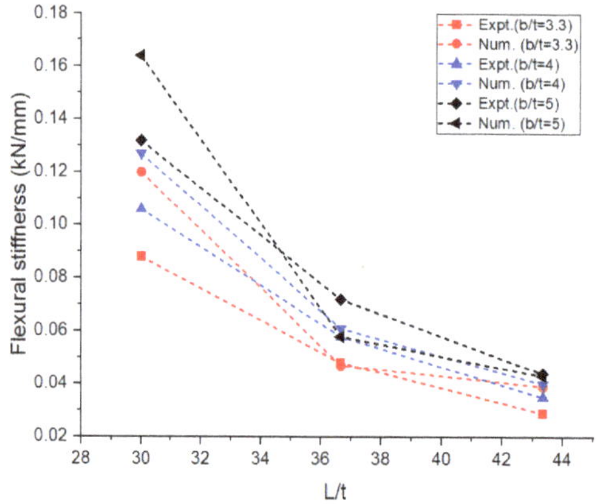

Figure 6. Aspect ratio vs. flexural stiffness: numerical and experimental.

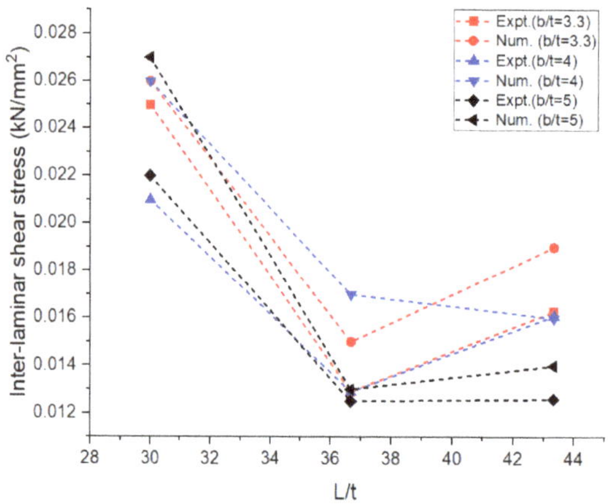

Figure 7. Aspect ratio vs. inter-laminar shear stress: numerical and experimental.

In the numerical analysis, the sandwich composite laminate consisting of high-density polyethylene core was assumed to be a homogeneous material with negligible defects. The maximum bending load was obtained at a lower L/t and a higher b/t, corresponding to 30 and 5, respectively. Improved longitudinal shear stress was the reason for this effect. At a higher L/t, the sandwich composite laminates became very weak against bending due to the accumulation of stress at the contact surfaces between the metal face layer and the core. The stress accumulated on the contact surfaces varied across the span of the laminate. This was confirmed from the observations of the modeled output of the maximum bending load in the range of 0.65 to 0.75 kN at L/t = 43.33 compared to that at L/t = 30, where it ranges between 1 to 1.6 kN. The corresponding maximum deflection complemented the output measure, i.e., the bending load. The deflection measured doubled (~98%) when the aspect ratio, L/t, increased from 30 to 43.33. Thus, the weakening effect of the sandwich laminates at a higher L/t resulted in increased deflection and poor flexural stiffness as well. The results obtained using both numerical and experimental approaches were complementary with respect to each other, with the least significance of error. The errors obtained in the performance measures between the experimental and numerical approaches were less than 10% in most of the trials except for a few. The flexural stiffness of about 0.164 kN/mm for L/t = 30 and b/t = 5 reduced drastically to 0.043 kN/mm for the same b/t and L/t = 43.33. A similar failure phenomenon was observed in the results of inter-laminar shear stress of the sandwich laminate. An almost 50% reduction was seen in the inter-laminar shear stress of the laminate with an increase in the aspect ratio, L/t, from 30 to 43.33. This once again confirmed that the process was consistent and accurate in terms of design and modelling remaining closer to the experimental conditions. The accuracy and precision of the results obtained reflect the fact that the same replications were carried out in the numerical study and experimentation. From the observations of the buckling test shown in Table 2 and Figure 8, it can be gleaned that both the approaches, namely, experimental and numerical, resulted in consistent and complementary data. The maximum critical buckling load was obtained at an aspect ratio, b_3/t, corresponding to the width of 10 mm of the sandwich specimen. Since the buckling load was compressive in nature, the critical buckling load capacity was greatly influenced by the cross-section of the laminate. Compared to the cross-section of the laminate, the influence of end conditions during the buckling test was less significant. The decrease in the width of the sandwich laminate significantly reduced the buckling load capacity. The greater the width of the laminate, the higher the shear force distribution across the thickness of the laminate, which, in turn, increases the resistance toward lateral deflection and improves buckling performance [24,25]. In the fixed–fixed end condition, the resistance towards buckling was higher compared to that of other end conditions for the same dimensions of the laminate. The constrained moment and reactive forces developed at the ends of the laminate resulted in weaker slenderness and stronger buckling. The equivalent system of forces for the buckling load tended to fall in the failure region for a minimum cross-sectional area. The results obtained through the experimental and numerical approaches significantly demonstrated the fact that the approximated numerical models for bending and buckling behaviors were precise enough, with consistent results and the least possibility of errors. The sample output of numerical analysis for bending and buckling was shown in Figures 9 and 10.

Table 2. Numerical vs. experimental buckling load analysis results.

S. No.	End Condition	b/t	Bucking Load (N)		
			Num.	Expt.	Error (%)
1	1	5	155	158	1.9
2	1	4	116.2	122	4.8
3	1	3.3	101.9	99.78	−2.1
4	2	5	342	346.4	1.3
5	2	4	221	219.73	−0.6
6	2	3.3	204.6	206.9	1.1
7	3	5	655.8	654.8	−0.2
8	3	4	524.8	526.77	0.4
9	3	3.3	422	424.7	0.6

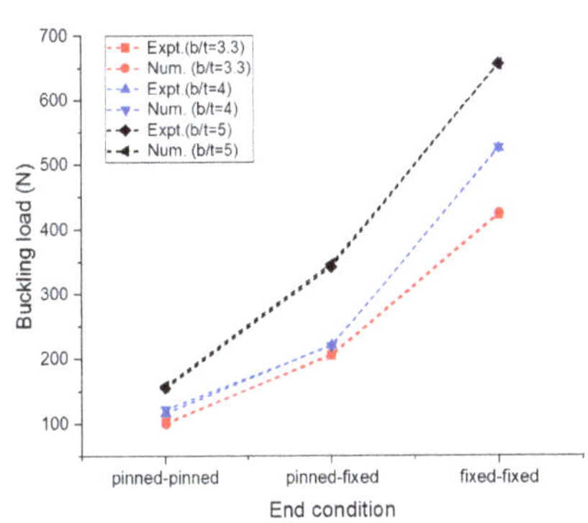

Figure 8. Effect of aspect ratio on buckling load: numerical and experimental results.

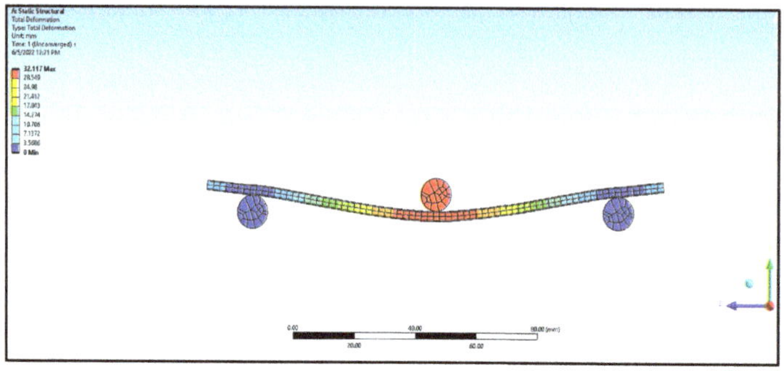

Figure 9. Sample ANSYS output for flexural test and buckling test.

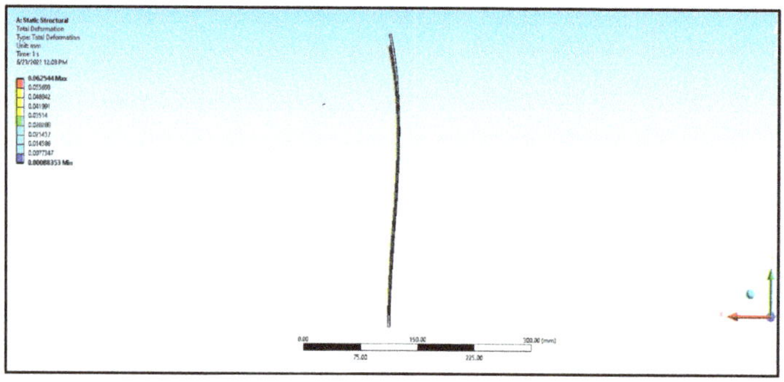

Figure 10. Sample ANSYS output for flexural test and buckling test.

4. Conclusions

Aluminum sandwich composite laminates (ASCs) were tested for their flexural and buckling behavior, and the following conclusions were drawn from the study.

(i) The influence of aspect ratios, i.e., the support-span-to-thickness and width-to-thickness ratios, on the flexural and buckling behavior of the sandwich composite laminate was significant.

(ii) The observations obtained from the flexural test revealed that the aspect ratios, L/t and b/t, influenced the laminate's flexural stability significantly. Though the adhesive layer connecting the metal face layer and the core contributed less to the bending behavior of the laminate, it significantly affected the overall ductility of the laminate.

(iii) Critical bending load and flexural stiffness were maximum for the support span with a width of 90 mm and 15 mm, corresponding to 3.6 kN and 4.75 kN/mm, respectively. The resistance offered against bending was maximum at a greater width. Similarly, with a higher support span, the spring-back effect was reduced, resulting in large-scale bending.

(iv) A higher magnitude of inter-laminar shear stress was noticed for the widths 10 mm and 15 mm, whereas it was minimum for the width of 12 mm. Hence, it was found that the optimum width of this sandwich laminate with a length of 150 mm was 12 mm in order to resist the delamination shear of the laminate.

(v) Maximum critical buckling load was obtained for the aspect ratio b_3/t, corresponding to the width of 10 mm of the sandwich specimen, where the contribution towards the buckling resistance was high.

(vi) The results obtained from the bending and buckling behavior of the aluminum–polyethylene sandwich laminate reveal that these kinds of panels perform better in design and stability for lower-altitude structures than for higher-elevated structures.

A lower aspect ratio (span to thickness) benefited these sandwich laminates to a greater degree in terms of both bending and buckling.

Author Contributions: Conceptualization, G.R. and D.B.; methodology, G.R., D.B and K.R.K.; software, K.R.K.; validation, G.R., D.B. and K.R.K.; formal analysis, A.H.M.S.A.H., A.M.N.A.Y., A.M.N.O.A.R. and A.M.D.A.H.; investigation, G.R., A.H.M.S.A.H., A.M.N.A.Y., A.M.N.O.A.R. and A.M.D.A.H.; resources, A.H.M.S.A.H., A.M.N.A.Y., A.M.N.O.A.R. and A.M.D.A.H.; data curation, G.R., D.B., A.H.M.S.A.H., A.M.N.A.Y., A.M.N.O.A.R. and A.M.D.A.H.; writing—original draft preparation, G.R., A.H.M.S.A.H., A.M.N.A.Y., A.M.N.O.A.R. and A.M.D.A.H.; writing—review and editing, K.R.K.; visualization, K.R.K. and D.B.; supervision, K.R.K. and D.B.; project administration, G.R.; funding acquisition to pay APC, D.B. All authors have read and agreed to the published version of the manuscript.

Funding: This research received no external funding.

Data Availability Statement: The data that support the findings of this study are available within the article.

Conflicts of Interest: The authors declare no conflict of interest.

References

1. Park, J.W.; Cho, J.U. Experiment and analysis of unidirectional CFRP with a hole and crank as sandwich-form inhomogeneous composite. *Adv. Compos. Mater.* **2019**, *28*, 103–114.
2. Mohd, A. Tensile Strength and Bonding in Compacts: A Comparison of Diametral Compression and Three- Point Bending for Plastically Deforming Materials. *Drug Dev. Ind. Pharm.* **2002**, *28*, 809–813.
3. Ali, I. Spring back behavior of fiber metal laminates with carbon fiber-reinforced core in V-bending process. *Arab. J. Sci. Eng.* **2020**, *45*, 9357–9366.
4. Ehsan, S. Innovative approach to mass production of fiber metal laminate sheets. *Mater. Manuf. Process.* **2018**, *33*, 552–563.
5. Joshua, T.O.; Alaneme, K.K.; Bodunrin, M.O.; Omotoyinbo, J.A. On the microstructure, mechanical behaviour and damping characteristics of Al-Zn based composites reinforced with martensitic stainless steel (410 L) and silicon carbide particulates. *Int. J. Lightweight Mater. Manuf.* **2022**, *5*, 279–288.
6. Dongyang, C. The effect of resin uptake on the flexural properties of compression molded sandwich composites. *Wind Energy* **2022**, *25*, 71–93.
7. Tang, E.; Zhang, X.; Han, Y. Experimental research on damage characteristics of CFRP/aluminum foam sandwich structure subjected to high velocity impact. *J. Mater. Res. Technol.* **2019**, *8*, 4620–4630. [CrossRef]
8. Berner, K.; Davies, J.M.; Helenius, A.; Heselius, L. The durability of structural sandwich elements. *Mater. Struct.* **1994**, *27*, 33–39. [CrossRef]
9. Latour, M.; D'Aniello, M.; Landolfo, R.; Rizzano, G. Experimental and numerical study of double-skin aluminium foam sandwich panels in bending. *Thin-Walled Struct.* **2021**, *164*, 107894. [CrossRef]
10. Khan, M.A. Experimental and numerical analysis of flexural and impact behavior off glass sandwich panel for automotive structural application. *Adv. Compos. Mater.* **2018**, *27*, 367–386. [CrossRef]
11. Mysmulski, P. Non-linear analysis of the postbuckling behavior of eccentrically compressed composite channel-section columns. *Compos. Struct.* **2023**, *305*, 116446. [CrossRef]
12. Abdullah, J.A. Experimental investigation of bond-slip behavior of aluminum plates adhesively bonded to concrete. *J. Adhes. Sci. Technol.* **2017**, *31*, 82–99. [CrossRef]
13. Stolbchenko, M.; Frolov, Y. The mechanical properties of rolled wire-reinforced aluminum composites at different strain values. *Mech. Adv. Mater. Struct.* **2020**, *27*, 1599–1608. [CrossRef]
14. Chauhan, S.; Sahu, S.; Ansari, M.Z. Effect of boundary support conditions of impact behavior of silicon pin reinforced polymer sandwich composite structure. *Adv. Compos. Mater.* **2020**, *41*, 5104–5115.
15. Xu, G.; Qin, K.; Yan, R.; Dong, Q. Research on failure modes and ultimate strength behavior of typical sandwich composite joints for ship structures. *Int. J. Nav. Archit. Ocean Eng.* **2022**, *14*, 100428. [CrossRef]
16. Lim, S.-S.; Wong, J.-Y.; Yip, C.-C.; Pang, J.-W. Flexural strength test on new profiled composite slab system. *Case Stud. Constr. Mater.* **2021**, *15*, e00638. [CrossRef]
17. Zhang, Z.; Abbas, E.M.; Wang, Y.; Yan, W.; Cai, X.; Yao, S.; Tang, R.; Cao, D.; Lu, W.; Ge, W. Experimental study on flexural behavior of the BFRC-concrete composite beams. *Case Stud. Constr. Mater.* **2021**, *15*, e00738. [CrossRef]
18. Fan, Y.; Yang, X.; He, J.; Sun, C.; Wang, S.; GU, Y.; Li, M. The variation mechanism of core pressure and its influence on the surface quality of honeycomb sandwich composite with thin face sheets. *J. Mater. Res. Technol.* **2021**, *15*, 6113–6124. [CrossRef]
19. Huang, Z.-C.; Zhang, Y.-K.; Lin, Y.-C.; Jiang, Y.-Q. Physical property and failure mechanism of self-piercing riveting joints between foam metal sandwich composite aluminum plate and aluminum alloy. *J. Mater. Res. Technol.* **2020**, *17*, 139–149. [CrossRef]

20. Ganesh, R.; Saravanan, M. Effect of fiber orientation on mechanical behavior of glass fiber reinforced polyethylene terephthalate foam sandwich composite. *Mater. Today Proc.* **2022**, *62*, 624–628.
21. Ganesh, R.; Al Hattali, A.H.; Al Yahyai, A.M.; Al Riyami, A.M.; Al Hadrami, A.M. Experimental study on the effect of aspect ratio on flexural behavior of Aluminium Sandwich Composite. *Eng. Technol. J.* **2022**, *40*, 1–6.
22. Davies, J.M. Sandwich panels. *Thin-Walled Struct.* **1993**, *16*, 179–198. [CrossRef]
23. Zniker, H. Energy absorption and damage characterization of GFRP laminated and PVC-foam sandwich composites under repeated impacts with reduced energies and quasi-static indentation. *Case Stud. Constr. Mater.* **2020**, *16*, e00844. [CrossRef]
24. Ou, Y.; Fernando, D.; Sriharan, J.; Gattas, J.M.; Zhang, S. A non- linear beam-spring-beam element for modelling the flexural behaviour of a timber-concrete sandwich panel with a cellular core. *Eng. Struct.* **2021**, *244*, 112785. [CrossRef]
25. Cheol-Won, K. Experimental strength of composites sandwich panels with cores made of aluminum honeycomb and foam. *Adv. Compos. Mater.* **2014**, *23*, 43–52.

materials

Article

Process Optimization, Microstructure and Mechanical Properties of Wire Arc Additive Manufacturing of Aluminum Alloy by Using DP-GMAW Based on Response Surface Method

Wenbo Du [1], Guorui Sun [2], Yue Li [3] and Chao Chen [2,*]

1 National Key Laboratory for Remanufacturing, Army Academy of Armored Forces, Beijing 100072, China
2 College of Mechanical and Electrical Engineering, Northeast Forestry University, Harbin 150040, China
3 Beijing Institute of Radio Measurement, Beijing 100039, China
* Correspondence: 1020210089@nefu.edu.cn; Tel.: +86-1884-3189-190; Fax: +86-0451-8219-0397

Abstract: Double-pulsed gas metal arc welding (DP-GMAW) is a high-performance welding method with low porosity and high frequency. Periodic shrinkage and expansion of the melt pool during DP-GMAW leads to unusual remelting, and the re-solidification behavior of the weld metal can significantly refine the weld structure. The advantages of DP-GMAW have been proven. In order to better apply DP-GMAW to aluminum alloy arc additive manufacturing, in this paper, the single-pass deposition layer parameters (double-pulse amplitude, double-pulse frequency and travel speed) of DP-GMAW will be optimized using the response surface method (RSM) with the width, height, and penetration of the deposition layer as the response values to find the superior process parameters applicable to the additive manufacturing of aluminum alloy DP-GMAW. The results show that the aluminum alloy components made by DP-GMAW additive are well formed. Due to the stirring of double-pulse arc and the abnormal remelting and solidification of metal, the microstructures in the middle and top areas show disordered growth. The average ultimate tensile strength of the transverse tensile specimen of the member can reach 175.2 MPa, and the elongation is 10.355%.

Keywords: double-pulsed gas metal arc welding; additive manufacturing; response surface method; parameter optimization; microstructure analysis; aluminum alloy

Citation: Du, W.; Sun, G.; Li, Y.; Chen, C. Process Optimization, Microstructure and Mechanical Properties of Wire Arc Additive Manufacturing of Aluminum Alloy by Using DP-GMAW Based on Response Surface Method. *Materials* **2023**, *16*, 5716. https://doi.org/10.3390/ma16165716

Academic Editor: Amir Mostafaei

Received: 14 November 2022
Revised: 3 December 2022
Accepted: 12 December 2022
Published: 21 August 2023

1. Introduction

Aluminum material is one of the important materials for lightweight design. Aluminum products have a large number of applications in the fields of aerospace, vehicles, and mechanical preparation [1–3]. Wire and Arc Additive Manufacturing (WAAM) is a typical process of Additive Manufacturing (AM) technology by digital means. Compared to traditional aluminum alloy manufacturing methods, such as casting, forging, and welding, WAAM has many advantages in terms of simplicity, freedom of design, and high material utilization [4–8].

Double-pulsed gas metal arc welding (DP-GMAW) is a high-performance welding method with low porosity, high frequency, and concentrated energy [9,10]. Numerous scholars have shown that the microstructure of DP-GMAW welds is significantly finer compared to pulsed gas metal arc welding (P-GMAW). The shear force at the peak of the strong pulse causes the dendrites to break up, providing enough nuclei for grain growth [11–13]. Based on the advantages of DP-GMAW, some scholars have studied DP-GMAW additive manufacturing and verified the feasibility of DP-GMAW additive manufacturing. Mainak Sen et al. [11] conducted overlay tests on mild steel plates using DP-GMAW with different combinations of parameters. The results show that the volume fraction of inclusions and needle ferrite in the weld metal increases with decreasing heat input, pulse frequency, and thermal pulse frequency. Yao P et al. [13] showed that the width, height, and depth of melt of the weld seam were positively correlated with the double-pulse relationship, average welding current, and percentage change in double-pulse current, and negatively correlated with travel speed and double-pulse frequency. Koushki A R et al. [14] added 0.1 vol% of oxygen and nitrogen to the shielding gas and the tensile and flexural strength of the weld was improved; however, the formation of oxide film was detrimental to the performance when the content was further increased.

The response surface method (RSM) is to fit the equation from the data of the experiment and represent it by means of a coordinate plot, which can predict the effect of different conditions on the response values. Moreover, the results can be optimized under specific conditions [15,16]. Haibin Geng et al. [17] developed a predictive model between the input variables (peak current, wire feed speed, and travel speed) and the response values (height and width of weld bead). The validity of the model was tested by analysis of variance (ANOVA). Waheed et al. [18] optimized the welding process parameters by RSM. Using optimal welding parameters optimizes the welding sequence. The results showed a 19% reduction in overall deformation caused by welding. Karganroudi et al. [19] used RSM to analyze the effect of current size and welding speed on weld geometry and temperature distribution. The penetration depth and penetration width decreased with the reduction in current residence time. Youheng F et al. [20] optimized the bainitic steel additive manufacturing process parameters using the RSM. The optimized specimen has a smooth surface with less splash and no visible defects. Escribano-García R et al. [21] combined RSM and finite element method (FEM) for 3D numerical simulation of GMAW cold metal transfer. They used RSM to find the optimal parameters. Sarathchandra D et al. [22] studied the effect of process parameters on WAAM of 304 stainless steel. RSM and ANOVA were used to evaluate the effects of current, travel speed, and weld distance on weld seam characteristics.

The advantages of DP-GMAW have been proven. In order to better apply DP-GMAW to aluminum alloy arc additive manufacturing, in this paper, the RSM was used to optimize the parameters of DP-GMAW additive manufacturing process, the effects of DP-GMAW process parameters on each response quantity of single-layer single-pass deposition layer of aluminum alloy were compared and analyzed, and a mathematical model was established between three process parameters of travel speed, double-pulse frequency, and double-pulse amplitude, and three response values of deposition layer width, height, and penetration. The regression model ANOVA of the width, height, and penetration of the deposition layer was also performed to check the significance of the model and the normal probability distribution of the model residuals. The relationship between the variables (travel speed, double-pulse frequency, double-pulse amplitude) and the response values (width, height, and penetration of the deposition layer) was analyzed using perturbation plots. A set of optimized parameters was selected for the deposition and forming of thin-walled components of ER4047 aluminum alloy, and their microstructure and mechanical properties were analyzed.

2. Materials and Methods

The Fast Mig X 350 (KEMPPI, Lahti, Finland) welder was selected for the test, and the current waveform was selected as a double-pulse waveform. The test system is shown in Figure 1a. The waveforms of typical welding currents during DP-GMAW [13] are shown in Figure 1b. The current waveform is a double-pulse cycle consisting of a strong pulse group (Pulse S) and a weak pulse group (Pulse W). Test material selection of 1.2 mm ER4047 aluminum alloy wire, substrate selection to remove the oxide layer of 2A12 aluminum alloy substrate, and wire and substrate composition are shown in Table 1. The protective gas (99.99% pure argon) flow rate was selected at 20 L/min, and the wire feed speed was fixed at 6 m/min. The traveling speed was precisely controlled by the single axis motion control box, and the speed value was measured and feedback by the sensor was installed on the electric ball screw system.

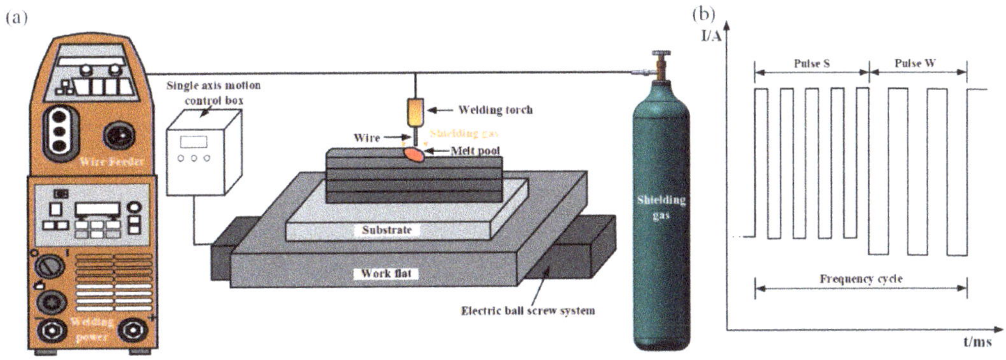

Figure 1. Diagram of the test system: (**a**) diagram of the Wire and Arc Additive Manufacturing (WAAM) process; (**b**) typical double pulse waveform diagram.

Table 1. Nominal compositions of ER4047 wire and 2A12 substrate (wt.%).

Element	Si	Fe	Cu	Mg	Mn	Ti	Zn	Al
ER4047	11~13	≤0.6	≤0.3	≤0.1	≤0.15	≤0.15	≤0.2	balance
2A12	≤0.50	0~0.5	3.8~4.9	1.2~1.8	0.30~0.9	≤0.15	≤0.30	balance

Box–Behnken design tests were selected using Design-Expert 10.0.7 software of StatEase company (Minneapolis, MN, America). Three key process parameters that control the morphology of the DP-GMAW additive deposition layer were selected as the study variables for the experiment, i.e., double-pulse frequency (F), double-pulse amplitude (A), and travel speed (V). The three key parameters of deposition layer width, height, and penetration were selected as response values. In order to investigate the fitting of the central area and ensure the repeatability of the test, the central point repeated test is set to five groups. In order to improve the arc additive forming rate, the range of process parameters was determined as shown in Table 2, based on a large number of preliminary experimental explorations, combined with the forming quality of the single-pass deposition layer.

Table 2. Input process parameters and working ranges.

Parameters (Unit)	Optimization Scope		
	−1	0	1
A (m/min)	0.3	1	1.7
F (Hz)	1	4	7
V (mm/s)	10	12	14

The RSM test parameters are shown in Table 3. Metallographic specimens were cut at the highest and lowest positions using a CNC wire cutter. After grinding with sandpaper and polishing with a metallographic polisher, they were etched using Keller's reagent. The cross-sectional morphology was photographed using a body microscope, and the values of single-pass deposition layer width, height, and fusion depth were measured separately. Since some of the parameters are fluctuations in the formation of the deposition layer surface, the average value of the two cross-sections was taken as the response quantity. Figure 2 shows the surface morphology of the deposition layer. The surface morphology of the deposition layer under different parameters is significantly different, as shown in Figure 2 (1–17).

Table 3. The response surface method (RSM) test parameters and corresponding response.

Std Order	Design Matrix			Input Variables			Responses		
				A: Double-Pulse Amplitude/(m/min)	B: Double-Pulse Frequency/(Hz)	C: Travel Speed/(mm/s)	Penetration P/(mm)	Width W/(mm)	Height H/(mm)
1	−1.00	−1.00	0.00	0.3	1	12	1.560	6.165	2.205
2	1.00	−1.00	0.00	1.7	1	12	1.525	7.150	2.085
3	−1.00	1.00	0.00	0.3	7	12	1.575	5.825	2.225
4	1.00	1.00	0.00	1.7	7	12	1.650	5.675	2.050
5	−1.00	0.00	−1.00	0.3	4	10	1.780	6.500	2.200
6	1.00	0.00	−1.00	1.7	4	10	1.800	7.025	1.775
7	−1.00	0.00	1.00	0.3	4	14	1.450	6.310	1.625
8	1.00	0.00	1.00	1.7	4	14	1.795	7.100	1.875
9	0.00	−1.00	−1.00	1	1	10	1.925	6.625	2.150
10	0.00	1.00	−1.00	1	7	10	1.550	6.100	1.505
11	0.00	−1.00	1.00	1	1	14	1.530	5.985	1.559
12	0.00	1.00	1.00	1	7	14	1.610	6.785	1.640
13	0.00	0.00	0.00	1	4	12	1.690	6.505	1.700
14	0.00	0.00	0.00	1	4	12	1.510	6.350	1.780
15	0.00	0.00	0.00	1	4	12	1.575	6.275	2.100
16	0.00	0.00	0.00	1	4	12	1.550	6.700	2.000
17	0.00	0.00	0.00	1	4	12	1.575	6.125	1.950

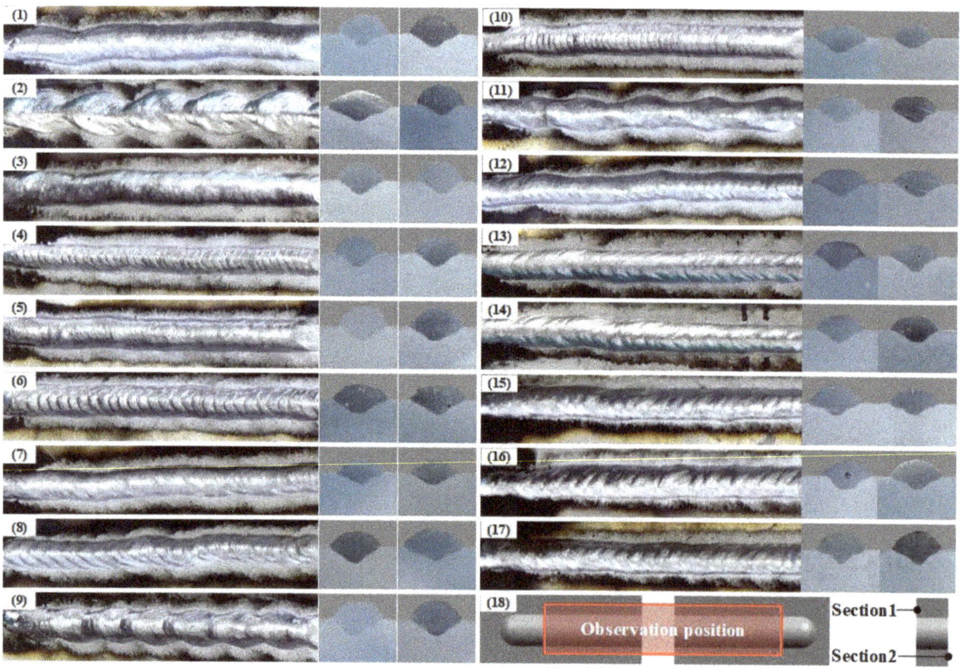

Figure 2. Macroscopic diagram of the deposition layer: (**1–17**) surface morphology and cross-sectional view of the deposition layer; (**18**) schematic diagram of sampling and shooting locations.

3. Results and Discussion

ANOVA was used to test the significance of the model and its misfit. The feasibility of the model is judged based on the magnitude of the Adeq Precision value. The functional relationship between the input variables and the response quantity can be expressed uniformly as y = f(A, F, V), expanding the response quantity y into the form of a second-order polynomial regression equation. After the calculation of the obtained coefficients, the functions between the deposition layer width, height, and penetration and the three input variables are shown in Equations (1)–(3), respectively. By analyzing the regression equation and combining the test results of the response in Table 3, the mathematical relationship between the input variables and the response within the parameter range is established to seek the optimal process parameters. The coefficients in the equation have been retained to three decimal places.

$$P = +1.57 + 0.051 \cdot A - 0.074 \cdot B - 0.084 \cdot C + 0.027 \cdot AB + 0.081 \cdot AC + 0.11 \cdot BC + 0.024 \cdot A^2 + 0.1 \cdot C^2 + 0.11 \cdot A^2B \quad (1)$$

$$W = +6.39 + 0.27 \cdot A + 0.069 \cdot B - 0.00875 \cdot C - 0.28 \cdot AB + 0.33 \cdot BC + 0.086 \cdot A^2 - 0.27 \cdot B^2 + 0.26 \cdot C^2 - 0.52 \cdot A^2B \quad (2)$$

$$H = +1.91 - 0.059 \cdot A - 0.072 \cdot B - 0.12 \cdot C - 0.014 \cdot AB + 0.17 \cdot AC + 0.18 \cdot BC + 0.20 \cdot A^2 + 0.040 \cdot B^2 - 0.23 \cdot C^2 \quad (3)$$

From Table 4, it can be seen that the model *F* value for penetration is 5.21. The value of probability $P > F$ is less than 0.05, indicating that the model is significant and the value of probability $P > F$ is 0.0203 (less than 0.05), so the model is significant and statistically significant. The coefficient of determination R-Squared(R^2) value of the fitted regression equation is 0.8701 (>0.80), which is relatively close to 1. These two points indicate that the fitted equation of the model is acceptable. The value of Lack of Fit (LF) is 1.22, which results in insignificant, indicating that the model is reliable. In determining the penetration depth, the influence of *C* travel speed is important, while *A* and *B* are not significant in determining the penetration depth. A signal-to-noise ratio greater than 4 is ideal, and the model's signal-to-noise ratio of 8.625 indicates that there is sufficient signal for the model to be used for prediction. According to the test results, when C value increases from 10 mm/s to 14 mm/s, the value of penetration *P* decreases gradually. This is due to the fact that the increase in travel speed will lead to a reduction in heat input per unit length and the reduction in the depth of molten substrate in the single factor change.

Table 4. Model analysis of variance (ANOVA) results for penetration.

Source	Sum of Squares	df	Mean Square	*F* Value	*p*-Value *P* Rob > *F*	Contribution (%)	
Model	0.23	9	0.026	5.21	0.0203		significant
A-DP-Amplitude	0.021	1	0.021	4.18	0.0802	8.281	
B-DP-Frequency	0.022	1	0.022	4.44	0.0732	8.796	
C-Travel Speed	0.056	1	0.056	11.44	0.0117	22.662	
AB	0.003025	1	0.003025	0.62	0.458	1.228	
AC	0.026	1	0.026	5.38	0.0534	10.658	
BC	0.052	1	0.052	10.55	0.0141	20.899	
A^2	0.002342	1	0.002342	0.48	0.5118	0.951	
C^2	0.042	1	0.042	8.57	0.0221	16.977	
A^2B	0.024	1	0.024	4.82	0.0641	9.548	
Residual	0.034	7	0.004905				
Lack of Fit	0.016	3	0.005462	1.22	0.4118		not significant
Pure Error	0.018	4	0.004488				
Cor Total	0.26	16					
Std. Dev.	0.07	R-Squared	0.8701				
PRESS	0.33	Adeq Precision	8.625				

As seen in Table 5, the model *F* value for width is 8.85, indicating that the model is significant. The value of LF is 0.34, which indicates that the model is reliable. Due to noise, there is a 79.90% probability that such a large "Lack of Fit *F*-value" will occur. The R^2 value of the coefficient of determination of the fitted regression equation is 0.9192 (greater than 0.80), which is closer to 1. A double-pulse amplitude has a significant impact on the width, while B double-pulse frequency and the C travel speed have no significant impact on the width. The signal-to-noise ratio of the model is 10.302, which indicates that the model has sufficient signal to be used for prediction. When A value rises from 0.3 m/min to 1.7 m/min, the overall heat input increases, resulting in a significant increase in the melting width of the substrate.

Table 5. Model ANOVA results for width.

Source	Sum of Squares	df	Mean Square	*F* Value	*p*-Value *P* Rob > *F*	Contribution (%)	
Model	2.78	9	0.31	8.85	0.0045		significant
A-DP-Amplitude	0.58	1	0.58	16.58	0.0047	22.8570	
B-DP-Frequency	0.019	1	0.019	0.54	0.4853	0.7444	
C-Travel Speed	0.0006125	1	0.0006125	0.018	0.8983	0.0248	
AB	0.32	1	0.32	9.24	0.0188	12.7382	
BC	0.44	1	0.44	12.6	0.0094	17.3702	
A^2	0.031	1	0.031	0.9	0.374	1.2407	
B^2	0.32	1	0.32	9.05	0.0197	12.4762	
C^2	0.28	1	0.28	7.94	0.0258	10.9460	
A^2B	0.55	1	0.55	15.67	0.0055	21.6025	
Residual	0.24	7	0.035				
Lack of Fit	0.05	3	0.017	0.34	0.799		not significant
Pure Error	0.19	4	0.049				
Cor Total	3.02	16					
Std. Dev.	0.19	R-Squared	0.9192				
PRESS	1.46	Adeq Precision	10.302				

The model *F*-value for height H is 4.25, which is significant as shown in Table 6. A "misfit *F*-value" of 0.49 means that the model is reliable. There is a 70.57% probability that such a large "out-of-fit *F*-value" will occur due to noise. The R^2 value of the coefficient of determination of the fitted regression equation is 0.8454 (greater than 0.80), which is closer to 1. The signal-to-noise ratio of the model is 6.952, indicating that there is sufficient signal for the model to be used for prediction. When the traveling speed is constant, the height of the deposition layer is mainly affected by the wire feeding speed. It can be seen from Table 6 that the influence of the amplitude of double pulse on the height of the deposition layer is not significant.

Table 6. Model ANOVA results for height.

Source	Sum of Squares	df	Mean Square	*F* Value	*p*-Value *P* Rob > *F*	Contribution (%)	
Model	0.8	9	0.089	4.25	0.0347		significant
A-DP-Amplitude	0.028	1	0.028	1.32	0.288	3.367	
B-DP-Frequency	0.042	1	0.042	2.01	0.1996	5.127	
C-Travel Speed	0.11	1	0.11	5.19	0.0569	13.238	
AB	0.0007563	1	0.0007563	0.036	0.8545	0.092	
AC	0.11	1	0.11	5.45	0.0522	13.901	
BC	0.13	1	0.13	6.31	0.0403	16.094	
A^2	0.16	1	0.16	7.68	0.0276	19.589	
B^2	0.006737	1	0.006737	0.32	0.5879	0.816	
C^2	0.23	1	0.23	10.89	0.0131	27.776	
Residual	0.15	7	0.021				
Lack of Fit	0.04	3	0.013	0.49	0.7057		not significant
Std. Dev.	0.14	R-Squared	0.8454				
PRESS	0.8	Adeq Precision	6.952				

Figure 3a–c shows the normal probability distributions of the residuals of the single-pass deposition layer width, height, and penetration models. The colors in Figure 3 represent responder values of different sizes. The residual distributions of all three models are approximately along a straight line, indicating that the regression models fit better, the

error distribution is more uniform, there are no singularities with large deviations, and the models can predict the response quantity values more accurately. Understanding the influence of each input variable (A, B, C) and its interaction terms (such as AB, AC, BC, etc.) on the nature of the response quantity can predict more accurately the trend of the deposition layer size with the change of the double-pulse process parameters.

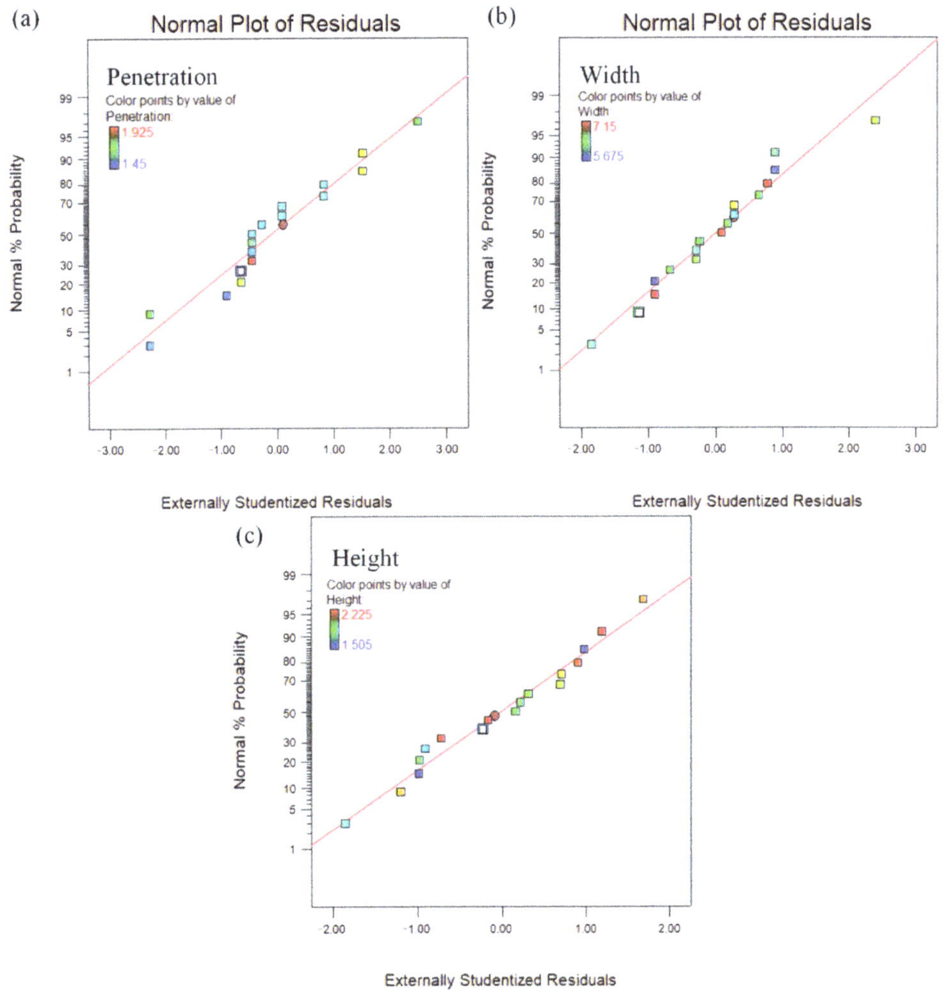

Figure 3. Normal plot of residuals for W, H and P: (**a**) penetration; (**b**) width; (**c**) height.

In order to deeply analyze the effects of the three input variables of double-pulse frequency (F), double-pulse amplitude (A) and travel speed (V) and their interaction terms on the response values, the perturbation plots of the deposition layer width (W), height (H), and penetration (P) were carefully studied, as shown in Figure 4a–c. From Figure 4a, it can be seen that travel speed has the largest amount of perturbation and the most significant effect on the penetration. When the value of the double-pulse amplitude deviates from the central reference point, the penetration of the deposition layer tends to increase gradually as travel speed decreases. This is due to the fact that the smaller the travel speed is, the greater the heat input per unit length and the greater the melting depth of the base material per unit area, i.e., the greater the melting depth. The width of the deposition layer is most significantly influenced by the double-pulse amplitude, as shown in Figure 4b. With the increase in double-pulse amplitude, the width increases significantly. In the DP-GMAW process, the double-pulse amplitude is regulated by the wire feed speed and increasing the double-pulse amplitude means increasing the wire feed speed in the pulse W phase. Therefore, width increases with the increase in double-pulse amplitude. The effects of double-pulse amplitude and travel speed on the deposition layer height are most obvious, and the effects of both on the deposition layer height show opposite trends, as shown in Figure 4c. The deposition layer height tends to decrease and then increase with the increase in double-pulse amplitude, and then increase and then decrease with the increase in travel speed. The deposition layer height is related to the ratio of WFS and travel speed [17] and is influenced by the interaction between the two.

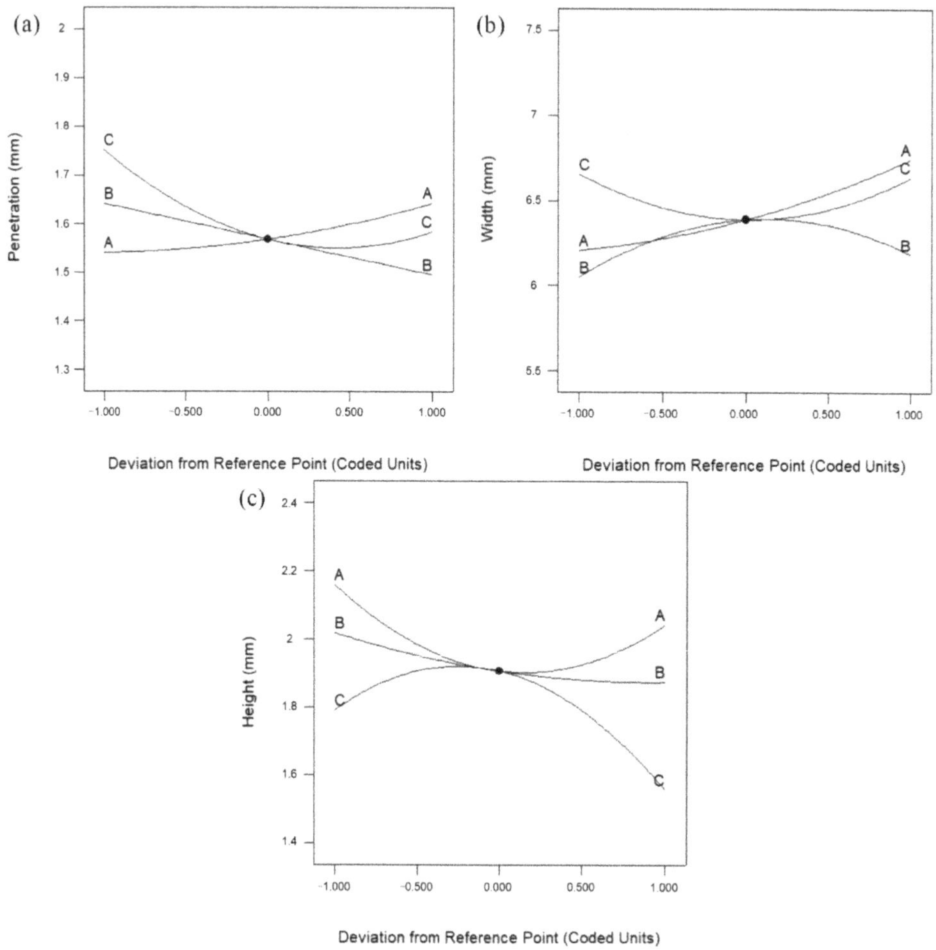

Figure 4. Perturbation diagrams of the deposition layer width, height, and penetration as a function of deviation of center reference point: (**a**) penetration; (**b**) width; (**c**) height. A: Double-Pulse Amplitude; B: Double-Pulse Frequency; C: Travel Speed.

After analyzing and verifying the reliability of the model, the DP-GMAW process parameters are optimized. In the DP-GMAW enrichment process, the double-pulse amplitude agitates the molten pool by modulating the pulse behavior, and the double-pulse frequency mainly controls the transition between the strong and weak pulse groups. The greater the frequency of the double pulse, the more frequent the transition between the strong and weak pulse groups, and the stronger the stirring effect on the molten pool. The grain size can be refined by the stirring effect of the electric arc [13,23]. Therefore, in the "Criteria" tab, set the double-pulse frequency f to Goal; maximize and the double-pulse amplitude to Goal; minimize, and select the scanning speed in the range of 10–14 mm/s as "in range". The feasibility index distribution of the optimized solution is obtained as shown in Figure 5, and the optimal parameters are obtained near the red area in the figure, and the feasible value of the solution is 1. The optimized results of this solution are double-pulse amplitude of 0.3 m/min, double-pulse frequency of 7 Hz, and travel speed of 12 mm/s.

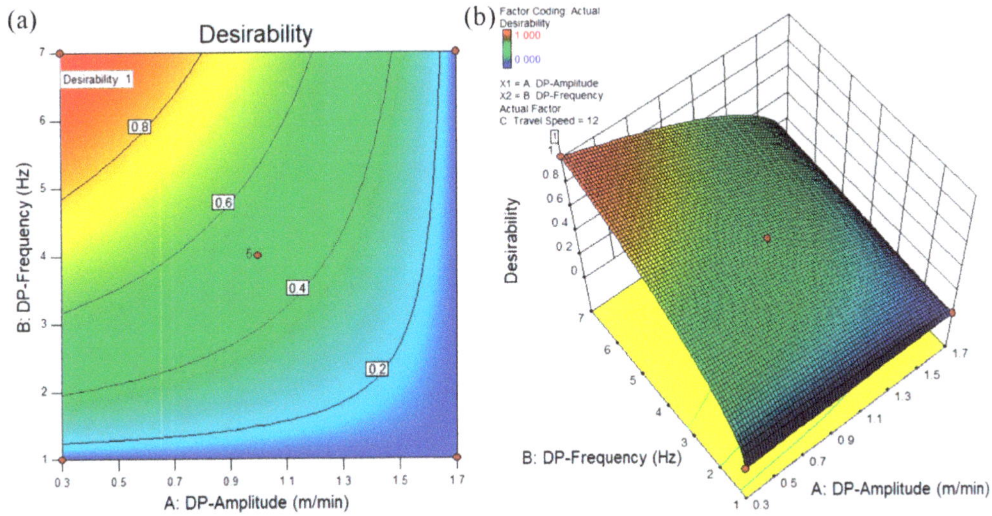

Figure 5. Effect of double-pulse amplitude and double-pulse frequency on the feasibility of the scheme with a travel speed of 12 mm/s: (**a**) contour plot of scheme feasibility; (**b**) 3D surface plot of probability distribution of scheme feasibility.

To investigate the effect of optimized parameters on DP-GMAW additive manufacturing, thin-walled Al-Si alloy parts with a height of 45 mm were prepared using the optimized parameters. The metallographic were cut at the positions shown in Figure 6b. The material is well connected between layers and no macroscopic defects, such as visible holes, cracks, and unfused layers, are found during or after deposition. The stirring action of the double pulse increases the melt pool width and reduces the layer height, measuring an effective wall width of about 8.6 mm. Figure 7 shows the microstructure of the specimen. There are differences between the interlayer and intra-layer microstructures, and the reason for the differences is that the interlayer microstructure has undergone remelting. The grains in the layer grow rapidly along the direction of maximum temperature gradient. The grains within the layer are mainly columnar crystals growing perpendicular to the fusion line, as shown in Figure 7c. Due to the stirring effect of the double pulses, under the influence of alternating periodic arc force, the columnar grains at some locations are interrupted, showing disordered growth of shorter columnar grains, as in Figure 7a,b.

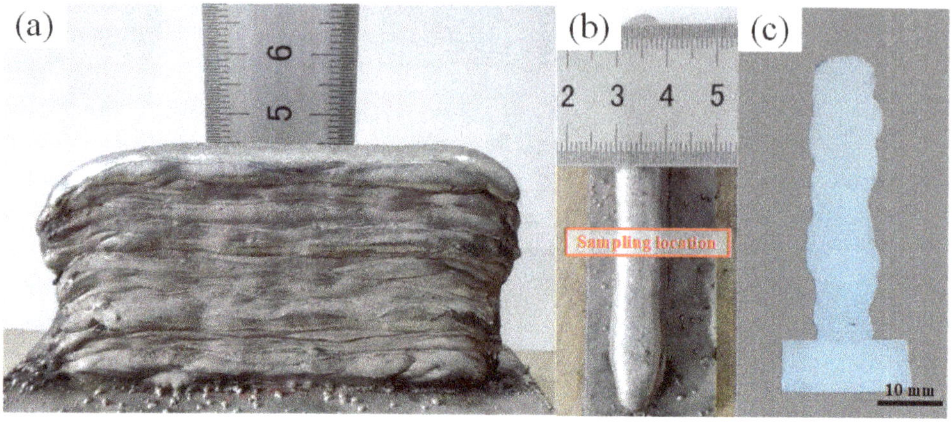

Figure 6. Shape of the WAAM components: (**a**) sidewall morphology; (**b**) wall width; (**c**) metallographic specimen.

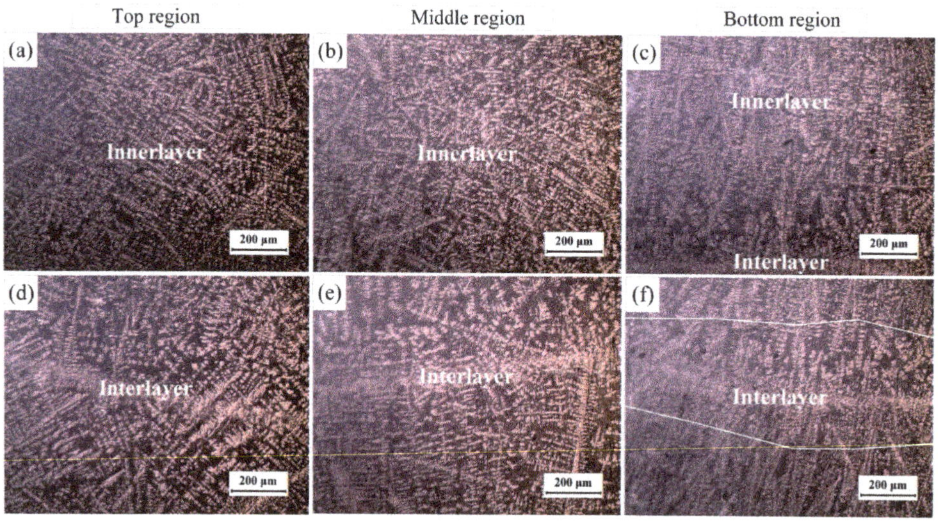

Figure 7. Microstructure at different positions: (**a**–**c**) intra-layer microstructure at the top and middle and bottom positions, respectively; (**d**–**f**) are the interlayer microstructures at the upper and middle positions, respectively.

There is less eutectic content between the layers. Due to the partial remelting effect of the arc on the deposition layer, part of the eutectic Si agglomerates in the molten pool are completely melted, and the other part adheres to the bottom of the pool to form large-size eutectic Si agglomerates, or the unmelted eutectic Si continues to grow. The eutectic Si in the interlayer microstructure is larger and sparser compared to the intra-layer size, as shown in Figure 7f. Figure 8 shows the EDS results, α-Al grains are mainly dendritic formations. Most silicon elements are concentrated in Al-Si eutectic, as shown in Figure 8d.

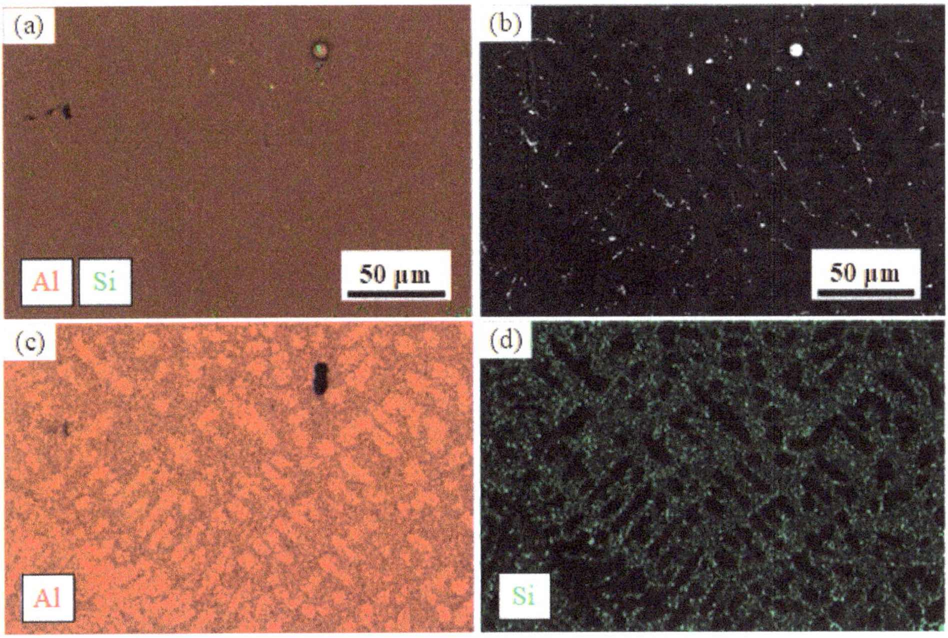

Figure 8. EDS results: (**a**) elemental layering images; (**b**) scanning electron microscope image; (**c**) Al element distribution; (**d**) Si element distribution.

The Al elements form α-Al dendrites except for the eutectic organization with Si elements. The α-Al dendrites in the bottom layer grow approximately parallel and long, while the α-Al dendrites in the middle and top regions grow in a disordered crossed state. In addition to the stirring effect of the double-pulse arc on the melt pool, the grain growth is also related to the heat dissipation of the melt pool. When the initial layer is deposited, the heat from the melt pool can be easily directed to the substrate for heat dissipation, and a small portion of the heat is dissipated through the air by convection and radiation, and the dendrites grow rapidly in the direction of the temperature gradient. When the top layer is deposited, most of the heat can only be radiated through the air by convection and radiation due to the large amount of heat accumulated in the previous deposition layers. Therefore, the growth pattern of α-Al dendrites at the top of the incremental body shows interlocking growth. When the middle layer is deposited, the temperature of the incremental body and the environment are relatively stable, the temperature gradient of the melt pool is small, and the α-Al dendrites grow along the vertical direction of the fusion line. Because of the stirring effect of the double-pulse arc, the α-Al dendrite growth is interrupted and shows a staggered growth pattern.

Figure 9 shows the drawing of tensile results, the sampling position of tensile specimens in transverse and vertical directions, the dimensions of tensile specimens, as shown in Figure 9a,b, and the ultimate tensile strength and elongation, as shown in Figure 9c. The results show that the ultimate tensile strength and elongation of the transverse tensile specimens are higher than those of the vertical ones. The average ultimate tensile strength of the transverse tensile specimens was 175.2 MPa and the average elongation was 10.355%. The average ultimate tensile strength of the vertical tensile specimens was 154.25 MPa and the average elongation was 7.46%. The mechanical properties exhibited significant anisotropy.

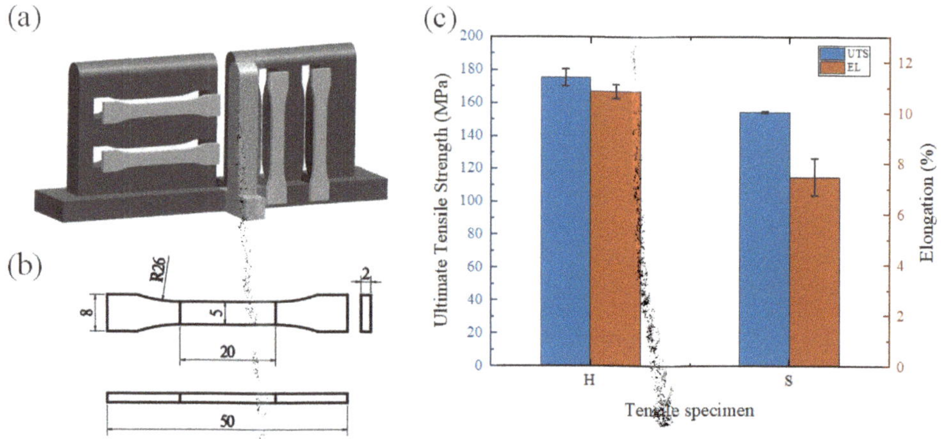

Figure 9. Drawing of tensile specimen: (**a**) sampling position of tensile specimen; (**b**) size of tensile specimen (mm); (**c**) ultimate tensile strength (UTS) and elongation (EL) of specimen. H—horizontal tensile specimen, S—vertical tensile specimen.

Figure 10 shows the SEM of the fracture of the tensile specimens. Air pores and a large number of dimples were found in both transverse and vertical tensile specimen fractograms. Dimples are a typical feature of ductile fracture. During the WAAM process, due to the high solubility in liquid aluminum and the high thermal conductivity and fast cooling of aluminum alloy, the hydrogen in the melt pool cannot escape in time and exists in the form of gas in the melt pool to form pores. The pores over 50 μm will have an impact on the mechanical properties of the components.

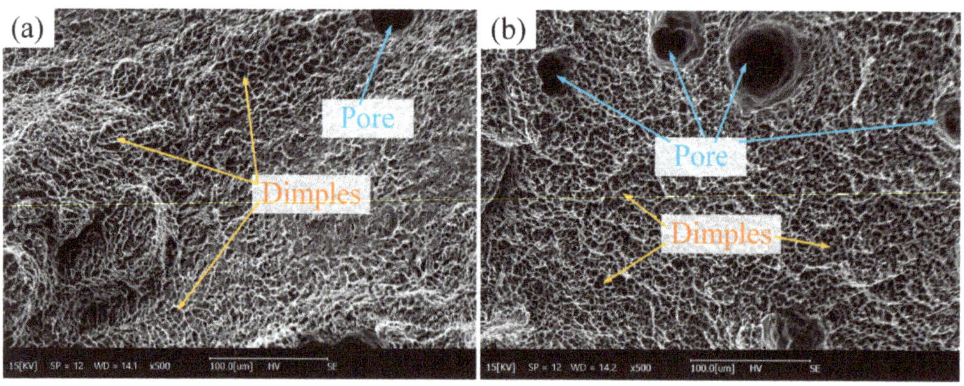

Figure 10. Fracture morphology (500×): (**a**) tensile specimen H; (**b**) tensile specimen S.

4. Conclusions

In order to better apply DP-GMAW to aluminum alloy arc additive manufacturing, the impact of three important parameters of DP-GMAW on the three evaluation indicators was analyzed by response surface method, and the following conclusions were drawn:

1. The models between the variables (travel speed, double-pulse frequency, and amplitude) and the response values (width, height, and fusion depth of the deposition layer) were significant. The models all had signal-to-noise ratios greater than 4, with adequate signal.

2. The residual distributions of all three models were approximately along a straight line and the regression models fitted well. The perturbation diagram shows that the penetration is most strongly perturbed by the travel speed, with a gradual increase in penetration as the travel speed decreases. The width of the layer is most significantly influenced by the double-pulse amplitude, while the height of the layer is influenced by the interaction between the double-pulse amplitude and the travel speed.
3. No macroscopic defects were observed during or after the deposition. The thin-walled parts are well formed with an effective wall width of 8.6 mm. Because of the stirring effect of the double-pulse arc, the growth of some α-Al dendrites within the layers is interrupted and shows a staggered growth pattern. The interlayer microstructure undergoes remelting and differs from the intra-layer microstructure.
4. The mechanical properties show a clear anisotropy. The average ultimate tensile strength of the transverse tensile specimens was 175.2 MPa and the average elongation was 10.355%. The average ultimate tensile strength of the vertical tensile specimens was 154.25 MPa and the average elongation was 7.46%.

Author Contributions: Design, C.C. and G.S.; investigation, W.D. and Y.L.; writing—original draft preparation, C.C. and G.S.; writing—review and editing, C.C. and G.S.; funding support, W.D. All authors have read and agreed to the published version of the manuscript.

Funding: This research received no external funding.

Institutional Review Board Statement: Not applicable.

Informed Consent Statement: Not applicable.

Data Availability Statement: Not applicable.

Conflicts of Interest: The authors declare no conflict of interest.

References

1. Zhang, J.; Song, B.; Wei, Q.; Bourell, D.; Shi, Y. A review of selective laser melting of aluminum alloys: Processing, microstructure, property and developing trends. *J. Mater. Sci. Technol.* **2019**, *35*, 270–284. [CrossRef]
2. Langelandsvik, G.; Akselsen, O.M.; Furu, T.; Roven, H.J. Review of Aluminum Alloy Development for Wire Arc Additive Manufacturing. *Materials* **2021**, *14*, 5370. [CrossRef] [PubMed]
3. Galy, C.; Le Guen, E.; Lacoste, E.; Arvieu, C. Main defects observed in aluminum alloy parts produced by SLM: From causes to consequences. *Addit. Manuf.* **2018**, *22*, 165–175. [CrossRef]
4. Wu, B.; Pan, Z.; Ding, D.; Cuiuri, D.; Li, H.; Xu, J.; Norrish, J. A review of the wire arc additive manufacturing of metals: Properties, defects and quality improvement. *J. Manuf. Process.* **2018**, *35*, 127–139. [CrossRef]
5. Tomar, B.; Shiva, S.; Nath, T. A review on wire arc additive manufacturing: Processing parameters, defects, quality improvement and recent advances. *Mater. Today Commun.* **2022**, *31*, 103739. [CrossRef]
6. Tawfik, M.M.; Nemat-Alla, M.M.; Dewidar, M.M. Enhancing the properties of aluminum alloys fabricated using wire plus arc additive manufacturing technique-A review. *J. Mater. Res. Technol.* **2021**, *13*, 754–768. [CrossRef]
7. Veiga, F.; Suarez, A.; Aldalur, E.; Artaza, T. Wire arc additive manufacturing of invar parts: Bead geometry and melt pool monitoring. *Measurement* **2022**, *189*, 8. [CrossRef]
8. Aldalur, E.; Suárez, A.; Veiga, F. Metal transfer modes for Wire Arc Additive Manufacturing Al-Mg alloys: Influence of heat input in microstructure and porosity. *J. Mater. Process. Technol.* **2021**, *297*, 13. [CrossRef]
9. Mathivanan, A.; Devakumaran, K.; Kumar, A.S. Comparative Study on Mechanical and Metallurgical Properties of AA6061 Aluminum Alloy Sheet Weld by Pulsed Current and Dual Pulse Gas Metal Arc Welding Processes. *Mater. Manuf. Process.* **2014**, *29*, 941–947. [CrossRef]
10. Wang, L.; Xue, J. Perspective on Double Pulsed Gas Metal Arc Welding. *Appl. Sci.* **2017**, *7*, 894. [CrossRef]
11. Sen, M.; Mukherjee, M.; Singh, S.K.; Pal, T.K. Effect of double-pulsed gas metal arc welding (DP-GMAW) process variables on microstructural constituents and hardness of low carbon steel weld deposits. *J. Manuf. Process.* **2018**, *31*, 424–439. [CrossRef]
12. Wang, L.L.; Wei, H.L.; Xue, J.X.; DebRoy, T. Special features of double pulsed gas metal arc welding. *J. Mater. Process. Technol.* **2018**, *251*, 369–375. [CrossRef]
13. Yao, P.; Tang, H.; Bin, K.; Chen, M.; Zhou, K. Influence of pulse frequency on weld bead formation and mechanical performance of 316L stainless steel by double-wire DP-GMAW process. *Int. J. Adv. Manuf. Technol.* **2022**, *121*, 3369–3383. [CrossRef]
14. Koushki, A.R.; Goodarzi, M.; Paidar, M. Influence of shielding gas on the mechanical and metallurgical properties of DP-GMA-welded 5083-H321 aluminum alloy. *Int. J. Miner. Metall. Mater.* **2016**, *23*, 1416–1426. [CrossRef]

15. Shim, D.-S. Effects of process parameters on additive manufacturing of aluminum porous materials and their optimization using response surface method. *J. Mater. Res. Technol.* **2021**, *15*, 119–134. [CrossRef]
16. Al-Ahmari, A.; Ashfaq, M.; Alfaify, A.; Abdo, B.; Alomar, A.; Dawud, A. Predicting surface quality of γ-TiAl produced by additive manufacturing process using response surface method. *J. Mech. Sci. Technol.* **2016**, *30*, 345–352. [CrossRef]
17. Geng, H.; Xiong, J.; Huang, D.; Lin, X.; Li, J. A prediction model of layer geometrical size in wire and arc additive manufacture using response surface methodology. *Int. J. Adv. Manuf. Technol.* **2015**, *93*, 175–186. [CrossRef]
18. Waheed, R.; Saeed, H.A.; Butt, S.U.; Anjum, B. Framework for Mitigation of Welding Induced Distortion through Response Surface Method and Reinforcement Learning. *Coatings* **2021**, *11*, 1127. [CrossRef]
19. Karganroudi, S.S.; Moradi, M.; Attar, M.A.; Rasouli, S.A.; Ghoreishi, M.; Lawrence, J.; Ibrahim, H. Experimental and Numerical Analysis on TIG Arc Welding of Stainless Steel Using RSM Approach. *Metals* **2021**, *11*, 19.
20. Fu, Y.; Wang, G.; Zhang, H.; Liang, L. Optimization of surface appearance for wire and arc additive manufacturing of Bainite steel. *Int. J. Adv. Manuf. Technol.* **2016**, *91*, 301–313. [CrossRef]
21. Escribano-García, R.; Rodriguez, N.; Zubiri, O.; Piccini, J.; Setien, I. 3D numerical simulation of GMAW Cold Metal Transfer using response surface methodology. *J. Manuf. Process.* **2022**, *76*, 656–665. [CrossRef]
22. Sarathchandra, D.T.; Davidson, M.J.; Visvanathan, G. Parameters effect on SS304 beads deposited by wire arc additive manufacturing. *Mater. Manuf. Process.* **2020**, *35*, 852–858. [CrossRef]
23. Wang, L.L.; Wei, H.L.; Xue, J.X.; DebRoy, T. A pathway to microstructural refinement through double pulsed gas metal arc welding. *Scr. Mater.* **2017**, *134*, 61–65. [CrossRef]

materials

MDPI

Article

Influence of Contour Scan Variation on Surface, Bulk and Mechanical Properties of LPBF-Processed AlSi7Mg0.6

Theresa Buchenau [1,*], Marc Amkreutz [1], Hauke Bruening [1] and Bernd Mayer [1,2]

1 Fraunhofer Institute for Manufacturing Technology and Advanced Materials, Wiener Straße 12, 28359 Bremen, Germany
2 Faculty of Production Engineering, University of Bremen, 28359 Bremen, Germany
* Correspondence: theresa.buchenau@ifam.fraunhofer.de; Tel.: +49-421-2246-7421

Abstract: Metal additive manufacturing technologies have great potential for future use in load-bearing aerospace applications, requiring a deeper understanding of mechanical performance and influencing factors. The objective of this study was to investigate the influence of contour scan variation on surface quality, tensile and fatigue strength for laser powder bed fusion samples made of AlSi7Mg0.6 material and to create high-quality as-built surfaces. The samples were produced with identical bulk and different contour scan parameters to accommodate the investigation of the impact of as-built surface texture on mechanical properties. The bulk quality was evaluated by density measurements according to Archimedes' principle and tensile testing. The surfaces were investigated using the optical fringe projection method, and surface quality was assessed by the areal surface texture parameters *Sa* (arithmetic mean height) and *Sk* (core height, derived from material ratio curve). Fatigue life was tested at different load levels, and the endurance limit was estimated based on a logarithmic-linear relation between number of cycles and stress. All samples were found to have a relative density of more than 99%. Surface conditions distinctive in *Sa* and *Sk* were successfully created. The resulting mean values of the ultimate tensile strength σ_{ult} are between 375 and 405 MPa for 7 different surface conditions. It was confirmed that the influence of contour scan variation on bulk quality is insignificant for the assessed samples. Regarding fatigue, one as-built condition was found to perform as well as surface post-processed parts and better than the as-cast material (compared to literature values). The fatigue strength at the endurance limit for 10^6 cycles is between 45 and 84 MPa for the three considered surface conditions.

Keywords: additive manufacturing; laser powder bed fusion; LPBF; PBF-LB; contour scan variation; mechanical testing; tensile strength; fatigue; AlSi7Mg0.6; surface quality; bulk quality; areal surface texture parameters

Citation: Buchenau, T.; Amkreutz, M.; Bruening, H.; Mayer, B. Influence of Contour Scan Variation on Surface, Bulk and Mechanical Properties of LPBF-Processed AlSi7Mg0.6. *Materials* **2023**, *16*, 3169. https://doi.org/10.3390/ma16083169

Academic Editors: Mostafa Hassani, Hongze Wang and Greta Lindwall

Received: 30 January 2023
Revised: 27 March 2023
Accepted: 13 April 2023
Published: 17 April 2023

1. Introduction

Additive manufacturing (AM) technologies, in particular laser powder bed fusion (LPBF), are of extraordinary interest to the aerospace industry. Advantages of these technologies include a large increase in geometrical freedom and potential savings of material and overall production cost [1–4].

It is also desirable to use AM technology in load-bearing applications, but standards for part certification and quality assurance are not yet established. Hence, there is currently still a restriction to non-critical parts in aerospace systems [5,6]. Part of the work done to gain an understanding of the process–material–property relations needed as a foundation for part qualification is summarized in Section 1.1. Section 1.2 explains the contribution of this work to that same understanding.

1.1. Mechanical Properties of LPBF-Processed AlSi Alloys

Different review papers have suggested that there is an extensive number of studies on LPBF processing of materials like Ti-64 [7–11], Inconel 718 [11–14] or 316L steel [15–19].

LPBF-processing of aluminium alloys, however, has only gained importance in recent years [20–25]. Aboulkhair et al. found that this is related to the particularly challenging properties of aluminium alloys and aluminium alloy powders for laser processing. The powders are generally characterized by low flowability, which impacts powder layer recoating, and are prone to oxidation, causing porosities. Moreover, the high reflectivity of the common LPBF process wavelength range, low laser absorption and high thermal conductivity result in a need for high laser power [21].

Nonetheless, LPBF processing of aluminium alloys is interesting, especially for lightweight construction applications, as they are lightweight, strong, corrosion-resistant and highly weldable. Combined with the geometrical freedom enabled by LPBF processing, they are suitable for tailoring parts for numerous purposes within automotive, aerospace and other industries [21].

The best LPBF-processable alloys are aluminium–silicon-based, and the most commonly-investigated one is AlSi10Mg. The silicon phase in the solidified LPBF material contributes to limit crack initiation and propagation due to the LPBF-typical fine microstructure and improves its tensile strength as compared to the cast material [4,24,26].

In particular, the alloys AlSi10Mg, AlSi12 and AlSi7Mg are considered 'highly printable' [24]. For these materials, ultimate tensile strength (UTS) values between 300 and 450 MPa in as-built condition are reported [4,24,27–36].

Many publications assess the effect of heat treatment [30,31,33,36–41], and there is some work addressing the effect of surface post-processing [42–45] or positioning on the build platform [39,46] on mechanical properties.

In this section, an overview of recent work on mechanical properties is given. The focus is on investigations on tensile and fatigue behavior of LPBF-processed aluminium alloys, particularly the AlSi7Mg0.6 alloy.

1.1.1. Tensile Properties

Yang et al. investigated the effect of heat treatments on microstructure and mechanical behavior anisotropy for the AlSi7Mg0.6 alloy. They observed the typical LPBF fine microstructure in as-built condition due to the material's fast cooling rate and a resulting higher strength than the as-cast alloy. Of the heat-treated samples, directly aged (T5) samples showed the highest strength and stress-relieved samples showed the largest elongation at fracture [31].

Similarly, Rao et al. found better tensile strength in as-built LPBF compared to the as-cast condition and observed that stress relaxation had a negative effect on yield strength (YS) and UTS while causing a slight improvement in ductility. A short solution heat treatment improved ductile behavior, and a longer treatment led to a decrease in YS and ductility [30].

Pereira et al. compared microstructure and mechanical properties of AlSi7Mg0.6 from LPBF and investment casting. They found that mechanical properties of LPBF can exceed aerospace qualification requirements for heat treated (T6) investment casting parts. They used direct aging heat treatment to improve ductility and hardness of LPBF-processed samples while maintaining a similar tensile strength as compared to as-built samples (e.g., mean UTS (as-built, vertical) of 435 MPa, after heat treatment 431 MPa) [33].

Zhang et al. looked into the effect of heat treatment for Er-containing AlSi7Mg0.6 and found that tensile properties are superior to the non-Er-containing alloy. The applied heat treatments improved ductility from 8% up to 19% for stress-relieved samples (with reduced tensile strength). Direct aging and T6 heat treatment both resulted in increased YS [36].

Advantages and disadvantages of different heat treatments compared to as-built ones were discussed by Mauduit et al. Amongst others, they found that the investigated heat treatments soft annealing and T6 resulted in isotropic mechanical properties. Soft annealing reduced tensile strength but removed residual stresses, artificial aging created the best UTS, but samples exhibited anisotropic mechanical properties. As-built samples already reached good mechanical properties but showed anisotropy. However, not applying heat treatment led to shorter production time and was less expensive [40].

Menezes et al. evaluated the effect of orientation on the build plate for as-built and heat-treated samples. Both conditions showed anisotropic behavior, where vertical specimens had lower YS and higher UTS. Comparing artificially aged and as-built samples, the latter showed lower YS [47].

Next to vertically (90°) and horizontally built (0°) samples, Denti included specimens built at a 45° angle in their investigation and observed a (slight) tendency for increasing tensile strength and decreasing elongation at fracture for steeper build angles [48].

In addition to heat treatment, Han et al. looked into the effect of laser surface remelting (LSR) for LPBF-processed AlSi10Mg and found that *Ra* (arithmetic mean of profile height variation) can be significantly improved by LSR. For as-built surfaces, they report an $Ra > 19$ µm that improved to values below 1 µm for LSR-processed samples. In addition, LSR led to increasing micro-hardness. The applied heat treatment led to reduced tensile strength and improved ductility from 6% to 22% [41].

1.1.2. Fatigue Properties

A full tension–tension loading Wöhler curve assessment with $R = 0$ of the AlSi7Mg0.6 alloy using an endurance limit of 2×10^6 cycles was performed by Bassoli et al. [49]. They obtained a result of 60 ± 5.3 MPa and found that the alloy's fatigue performance under the applied processing conditions was slightly lower but still comparable to reported literature values for the AlSi10Mg alloy [50]. Surface texture parameters were not specified, but they mentioned that the samples had not received any post-treatment.

Grande et al. [39] investigated the relationship of heat treatment and tensile strength as well as the effect of position on the build platform on fatigue life. They produced specimens with densities > 98.8% and as-built YS of 222 MPa and UTS of 417 MPa. They found that stress relief reduced tensile performance. Their fatigue results suggest that the position on the build platform does not have a significant influence on the endurance limit (at 10^7 cycles: 127 MPa internal vs. 137 MPa external regions) of the heat-treated specimens. Fatigue samples were sandblasted to improve surface texture prior to fatigue testing.

Denti and Sola [43] looked into the effect of different post-processing technologies (e.g., sandblasting, plastic media blasting and laser shock processing) on axial fatigue. They found that the evaluated surface processing techniques improved the areal arithmetic mean surface height deviation *Sa* by up to 77%. The lowest *Sa* values were achieved by plastic media blasting. The peak stress level at the endurance limit of 2×10^6 was improved by up to 80% with respect to the as-built σ_{max} of 50 MPa. Fatigue performance was also improved by post-processing techniques not enhancing the surface quality, which led them to the conclusion that both the improvement of surface quality and the introduction of compressive residual stresses can play a role when looking at LPBF-processed aluminium alloy parts.

The impact of sample location on the build platform, orientation and variation between production batches was studied by Cacace et al. [46]. By analyzing mechanical property data of three batches with randomly allocated sample positions, they found that part position did not have an influence on tensile strength but did affect low cycle fatigue performance.

Nasab et al. [51] investigated the combined effect of volumetric and surface defects. They looked into as-built surfaces with different contour scans, trying to promote typical defects to show their effect on rotating bending fatigue. The defect depths were analyzed by optical line-of-sight measurement, as well as polished cross-sections. Material removal depths of up to 200 µm were suggested, depending on the surface condition. They state that contact and non-contact surface texture measurements cannot provide information on fatigue-critical surface features as comprehensively as investigations into polished cross-sections.

In previously published work, the authors of this paper evaluated crack initiation behavior and surface fatigue relations for AlSi7Mg0.6 for three different groups of as-built samples. We assessed the applicability of valley depth *Sv* and reduced valley depth *Svk* and found that *Svk* is especially useful when considering coarser as-built surfaces, since they tend to exhibit crack initiation from multiple surface defects [52].

1.2. Motivation and Objective

Most of the studies summarized deal with the influence of heat treatment and build direction (horizontal/vertical) on tensile properties. While tensile properties are an important starting point in understanding a material's mechanical behavior and are certainly relevant for various applications, for many aerospace, automotive, biomedical or other industrial purposes, resistance to periodic loading is of interest. In regard to fatigue life, surface texture plays an important role [4,6].

The majority of studies including the effect of surface condition on fatigue performance of the LPBF-processed AlSi7Mg0.6 material, as well as other aluminium alloys and other typical LPBF powder materials (e.g., Ti-64, 316L steel or Inconel 718), evaluate the application of different surface post-processing strategies, e.g., [7,10,15,42,45,53–55].

However, especially when considering complex geometries or parts with inner surfaces that are difficult to access with post-processing tools, it is desirable to produce as-built surfaces (including near-surface regions) good enough to perform reasonably well under cyclic loading. In addition to accessibility issues, using as-built parts saves time and cost due to reducing processing steps, since extensive post-processing becomes unnecessary. In this paper, the effect of the as-built surface condition on mechanical properties is discussed. The ultimate aim is to create high-quality as-built surfaces.

The first step is to create distinctive as-built surface conditions by varying contour scan parameters (Section 3.1). Afterwards, the effect of these variations on bulk quality, characterized by density (Section 3.2) and UTS (Section 4.1), is investigated. Finally, a first selection of fatigue results is presented, showing the influence of as-built surface condition on fatigue resistance at a load level of $0.5\sigma_{ult}$ and the endurance limit (Section 4.2).

2. Materials and Methods

2.1. Manufacturing

The evaluated samples were manufactured in an LPBF process on a Trumpf TruePrint 1000 from AlSi7Mg0.6 aluminium alloy powder. The powder composition along with mass fractions of alloying elements are shown in Table 1. Specifications of geometries and manufacturing settings are given subsequently.

Table 1. AlSi7Mg0.6 powder composition: mass fraction per alloying element.

Al	Si	Mg	Ti	Fe
93.13	6.15	0.6	0.09	0.05

2.1.1. Sample Geometry

Two kinds of samples are used in this work: cuboids (height 10 mm, width 10 mm, thickness 5 mm) and fatigue specimens according to ASTM 466-15 [56] (height 80 mm, smallest cross section 6 mm, thickness 3 mm). The latter type is shown in Figure 1.

2.1.2. Manufacturing Parameters

Detailed information on the manufacturing process is presented in Tables 2 and 3 as well as Figures 1 and 2.

Powder layers were exposed to the laser by a pattern of parallel lines in the bulk, changing direction by 66° after each layer, and a continuous scan of the geometric contour. Sky writing was applied to ensure the laser source was moving at the chosen speed prior to exposure.

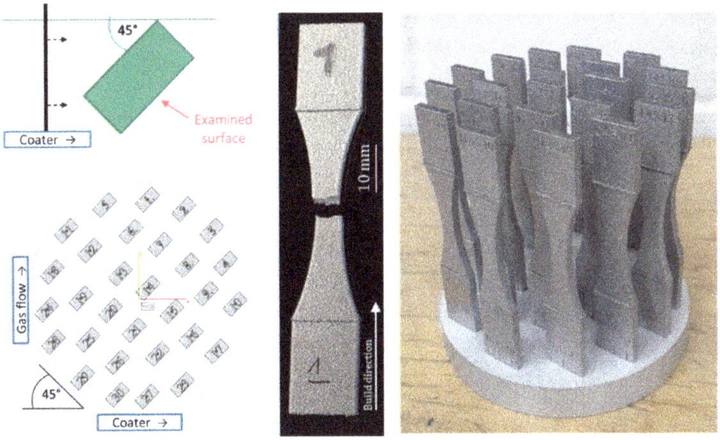

Figure 1. Top view: sample orientation with respect to coater (top left); example of build job layout (bottom left); individual tested sample (middle); finished build job on platform (right).

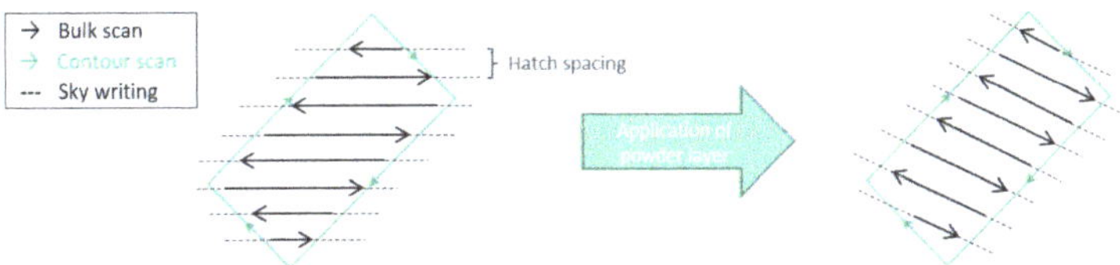

Figure 2. Exposure strategy for bulk and contour scan: bulk scan direction is rotated by 66° (schematic representation, not true to scale) after each powder layer application.

The samples were placed on the build platform at a 45° angle with respect to the coater and gas flow, as shown in Figure 1. This angle was found to be most suitable regarding surface texture. In preliminary studies, comparable surface texture parameter values were found for both sides of the sample, supposedly because the effects of coater and gas flow compensate each other.

Bulk scan parameters were identical for all samples, as specified in Table 2, and originate from a previous density optimization study.

The contour scan parameters were varied, intending to achieve a variation of surface properties. Maintaining layer thickness, hatch distance and laser power, the scan speed was modified between 300 mm/s and 1800 mm/s, paired with the settings with and without additional pre-sinter at 50% laser power, resulting in a total of 10 manufacturing parameter combinations.

The samples with identical parameter combinations were named with a designated letter according to Table 4, with consecutive numbering; e.g., A1 → Contour parameter set A (scan speed 300 mm/s, with pre-sinter), mechanical testing sample No. 1.

Table 2. Bulk scan parameters.

Material	Layer Thickness	Hatch Distance	Scan Speed	Laser Power	Pre-Sinter
AlSi7Mg0.6	30 µm	0.12 mm	1000 mm/s	195 W	No

Table 3. Variation of contour scan parameters.

Material	Layer Thickness	Hatch Distance	Scan Speed	Laser Power	Pre-Sinter
AlSi7Mg0.6	30 µm	0.12 mm	300 mm/s 600 mm/s 900 mm/s 1200 mm/s 1800 mm/s	195 W	Yes No

Table 4. Naming of sample groups based on contour scan variation.

Scan Speed in mm/s	300	600	900	1200	1800
Pre-Sinter	A	C	E	G	I
No Pre-Sinter	B	D	F	H	J

2.2. Characterization and Testing

2.2.1. Surface Texture

The surfaces were measured using a Keyence VR3200 fringe projection system. The micro camera setting at a magnification of 40× was applied, resulting in a lateral resolution of 7.4 µm. For the cuboid samples, selected ISO 25178 areal parameters were evaluated for a square area with an 8 mm length, measured perpendicular to the build direction on the side facing away from the coater, as indicated in Figure 1. A linear level operation, an S-filter of 20 µm and an L-filter of 0.25 mm were applied.

The chosen areal surface texture parameters to assess surface quality are *Sa*, the arithmetic mean height, and *Sk*, the core height from the material ratio curve. *Sa* was selected due to its common use in research and industry [57]. *Sk* is used because it gives more distinctive information on the surface texture (for details, refer to [58], p. 56).

The surface fatigue relation is shown using the material ratio curve parameter *Svk*, which is the reduced valley depth. The parameter was chosen because it describes the size of the valley population on the considered surface, rather than individual extreme values such as the maximum height *Sz* and the maximum valley depth *Sv*. More details can be found in [52]. *Sa*, *Sk* and *Svk* are defined in the ISO 25178-2 standard [59].

2.2.2. Density

The first step toward the assessment of bulk quality was the measurement of part density. For this purpose, the cuboid samples were weighed in air and ethanol using the Mettler Toledo Delta Range XS603S precision balance. The density was calculated according to Archimedes' principle as specified in ISO 3369 [60]. Each measurement was performed three times, and the final density result reported per sample is the respective mean value.

2.2.3. Tensile Testing

The tensile strength was tested using a ZWICK/Z050 in accordance with ASTM E8M [61]. A preloading of 35 N and a speed setting of 0.48 mm/min were selected.

The required cross-sectional areas of the tested specimens were obtained from digital caliper measurements.

2.2.4. Fatigue Testing

Fatigue life was tested on a DYNA-MESS 4S 20kN Z/D system at a frequency of 20 Hz and a stress ratio $R = 0.1$. The load levels were defined with respect to the mean value of the UTS for the tested surface conditions, $\sigma_{ult,mean} = 392$ MPa. Corresponding values are specified in Table 5.

Table 5. Load levels and stress values for σ_{ult} = 392 MPa and R = 0.1 [52].

Load Level $\sigma_{max}/\sigma_{ult}$	σ_{max}/MPa	σ_{min}/MPa	σ_{mean}/MPa
0.4	156.8	15.7	86.2
0.5	196.0	19.6	107.8
0.6	235.2	23.5	129.4
0.7	274.4	27.4	150.9

2.3. Workflow Summary

Figure 3 gives an overview of this work's process steps.

At the first manufacturing stage, 30 cuboid samples were made. All of these were manufactured with identical bulk scan parameters, paired with 10 variations of contour scan parameters, resulting in 3 cuboid samples per parameter set combination.

Afterwards, the cuboids' densities and surfaces were measured in order to get a first assessment of bulk quality and a rating based on the surface quality. Based on these evaluations, parameter sets were selected to produce samples for mechanical testing.

For seven manufacturing parameter sets chosen based on the cuboid assessment, six samples each were made for tensile testing. Tensile testing according ASTM E8M [61] was performed.

Finally, fatigue life was tested for a first selection of contour parameter sets, and their relationship with surface texture is discussed.

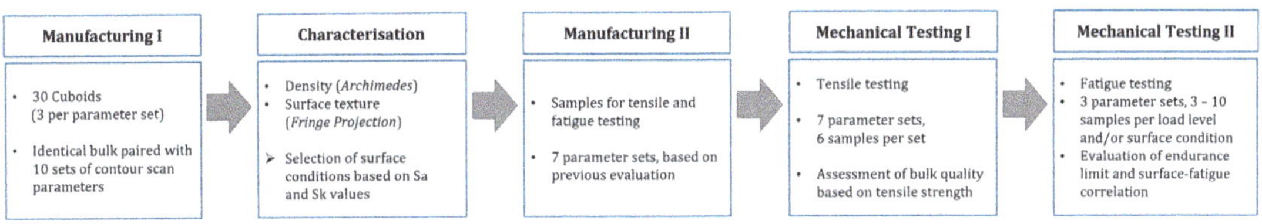

Figure 3. Workflow summary.

3. Results and Discussion of Preliminary Findings

The results presented in this section comprise density and surface texture characterization of the cuboid samples. The outcome is a selection of contour scan parameter sets for manufacturing the specimens for mechanical testing.

3.1. Surface Texture

3.1.1. Visual Perception of Surface Quality

From visual inspection of the microscopic images in Figure 4, it can be observed that, at first sight, a variety of as-built surface conditions was achieved.

The A and B conditions look mostly smooth with small dots and few linear defects (length below 1 mm, oriented parallel to the layers). Increasing the contour scan speed, surfaces appear to have more and bulkier linear defects (C and D). The D image also seems a little blurry, which is a sign of increasing height variation on the surface. This effect becomes more clear when increasing scan speed even further (E and F). On surface F, there are a few circular shadows present, which may be spatter or local accumulations of powder particles. Surfaces G to J are hardly distinguishable visually. All show circular shadows of different sizes, which are mostly particle agglomerations and accumulations, and an underlying irregular structure. Surface G shows some darker areas, which may be an issue of different lighting conditions or height differences on the surface itself.

With increasing scan speed, the energy absorbed by the powder in the scanned path decreases. Due to the low energy, powder particles are only partially molten and attached to the surface, causing coarse surface quality.

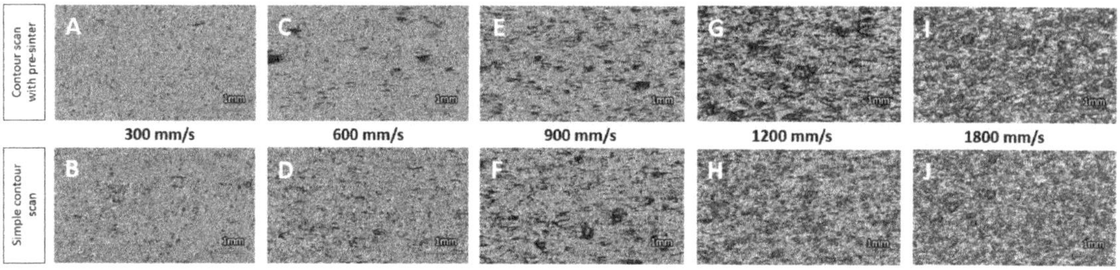

Figure 4. Microscopic images of samples with variation of contour scan speed, from lowest (left) to highest (right).

3.1.2. Selection of Contour Variation for Mechanical Testing from Surface Texture

Figure 5 shows *Sa* (left) and *Sk* (right) values. The surface conditions are sorted by contour scan parameters. To the right, results for sample sets exposed to pre-sinter are presented, while to the left of each graph, results for simple contour scans are shown. The scan speed increases from the middle to the edge.

The graphs give the mean (blue line) $\pm$ two standard deviations (SD, dashed blue line). Colors mark the surface conditions that are distinctive per a 95% confidence interval ($\pm$2SD) applied to the parameter results for *Sa* and *Sk*. The first group (red) includes surface conditions A to D; conditions E and F form the second group (green); conditions G and H (purple) are the third group; and finally, the fourth group (orange) comprises conditions I and J.

A superficial look at the graphs presented in Figure 5 already confirms that the objective of creating surfaces with varying surface quality was met. This is also supported by the microscopic images in Figure 4. Numerical values are included in Table A1.

Based on the graphs, parameter sets to produce specimens for mechanical testing were selected.

From the smooth (red) group including surface conditions A to D, A was chosen as the set with the lowest mean values for *Sa* and *Sk*. C and D were selected to compare the possible impact of pre-sinter with otherwise identical process settings (see Table 4). G and H from the purple group are both considered for the same reason as conditions with higher parameter values.

Conditions E and F (green group) show comparable mean values for *Sa* and *Sk*, and it was decided to use set E, as it was the original starting parameter set of the contour variation study, and to discard condition F.

Parameter set J is chosen as the set with the highest mean value for *Sa*. Condition I, as the second coarse texture set (orange group), has a larger SD for both considered surface texture parameters and was discarded.

In summary, the following contour parameter sets are applied to produce the specimens for mechanical testing:

- Smooth surface parameter sets A, C and D (red group),
- Original parameter set E (green group),
- Coarse (purple group) and very coarse (orange group) parameter sets G, H and J.

3.2. Density

The data shown in Figure 6 confirm a density of over 99% for all of the measured samples, denoted by the grey squares in the graph, with a reference density of 2.68 g/cm^3 (theoretical maximum).

When taking a 95% confidence interval (2SD), denoted by the dashed blue whiskers, into account, data sets A and B are below that 99% value. Data set A has a lower boundary value at 98.84%, which is also the lowest overall value.

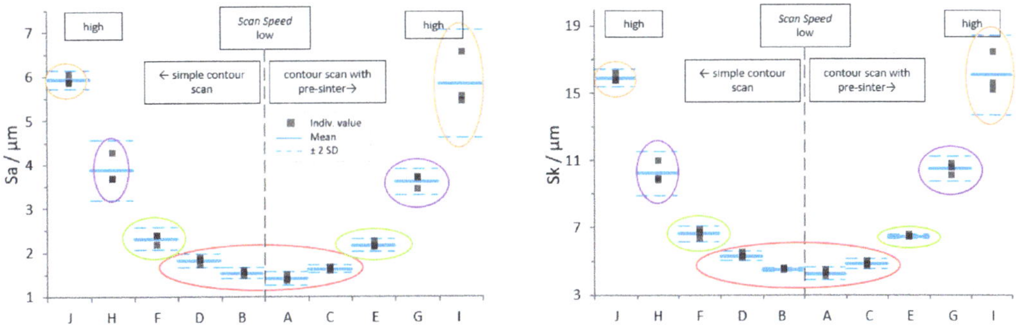

Figure 5. *Sa* and *Sk* for samples with different contour scan. L-filter 0.25 mm, S-filter 20 µm. Mean ± 2SD.

From all data sets, the only statistically distinctive sets considering the depicted 95% confidence interval (2SD) are B and G. However, they cannot be distinguished from the remaining data sets by that requirement.

In Figure 6, an increasing tendency of density for higher scan speed (A—lowest scan speed to J—highest scan speed) is observed. Supposedly, this is caused by the occurrence of closed porosities that can not be filled with ethanol during weighing. Possibly, the close proximity of the cuboids on the build platform during production plays a role as well, as the trend cannot be observed in the density data of the mechanical testing samples included in Table A2.

However, the overall mean and %SD including all 90 measured values (3 samples each for 10 contour parameter sets, 3 measurements each) are 99.5% and 0.3%, respectively. The mean and %SD taking individual groups A to J into account amount to 99.5% and 0.17%, respectively.

In conclusion, the evaluated sample sets are considered comparable. It is found from the presented results that the bulk scan parameters predominantly define the part density. Thus, varying contour scan parameters has no statistically significant influence on the density.

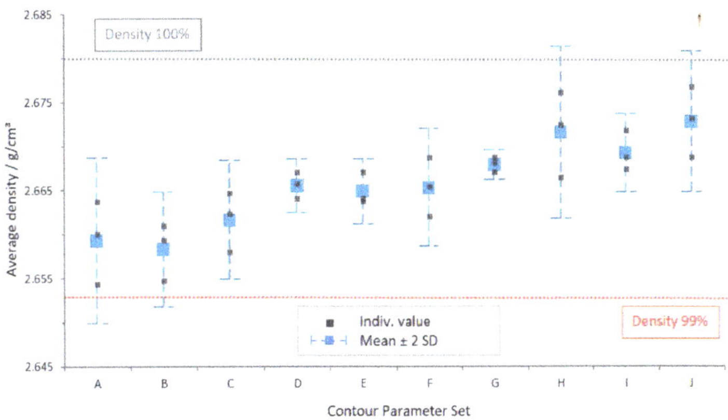

Figure 6. Density per manufacturing parameter set, mean ± 2SD; reference density: 100% = 2.68 g/cm^3.

4. Results and Discussion of Mechanical Characterization

4.1. Tensile Properties

Figure 7 shows the tensile testing results for longitudinal specimens manufactured using the contour parameter sets A, C, D, E, G, H and J. Density values for the mechanical testing samples, along with all numerical data presented in Figure 7, are included in Appendix A.

Similar to the results from density determination, there is no statistically significant difference in UTS. The mean values of the individual surface conditions are between 374 and 406 MPa. For context, values reported in the literature for as-built of the same material and build direction vary from 300 MPa [49] to over 400 MPa [30,39,40]. For the cast alloy with T6 heat treatment, typically values of UTS between 320 and 360 MPa are reported [30]. Hence, the tested samples perform equally well or better than other as-built LPBF AlSi7Mg0.6 specimens and mostly exceed the strength of the cast material.

From the graph, it can be observed that the standard deviation increases for rougher surface textures. A possible influencing factor is the caliper cross-section measurement, since the UTS depends on the cross-sectional area. The caliper may be locked by protruding features, leading to variation in measured cross-section.

Moreover, the combination of line energy and powder application is a potential explanation. Poor flowability properties affect the homogeneity of powder dispersion within a layer. At higher contour scan speeds, the high reflectivity and fast heat dissipation may lead to irregular density of molten material, causing coarser surface texture and different microstructural properties. The latter will have to be confirmed by a microstructural analysis.

The overall SD of UTS values, including all 40 test results, is low—3.8% (14.94 MPa). The SD within each group (1.3% to 4.4%) is of the same order of magnitude as the SD between the groups (2.5% between mean values). Hence, the sample groups produced with different contour scan parameters are considered comparable with regard to tensile strength.

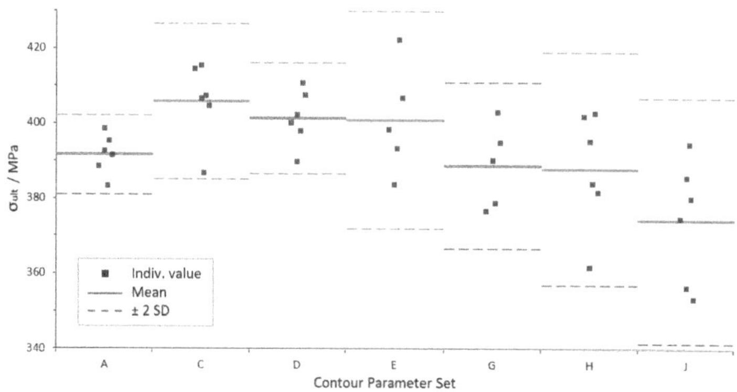

Figure 7. Ultimate tensile strength for seven different surface conditions, mean ± 2SD.

4.2. Fatigue Properties

The fatigue testing results for surface conditions A, E and G for load levels $\sigma_{max}/\sigma_{ult}$ 0.4, 0.5, 0.6 and 0.7 are presented in Figure 8. The smoothest surface condition, A, has the best fatigue performance for all load levels and low scatter, as expected. Even for the highest tested load level, cycle numbers above 10^4 are reached. Surface condition E exhibits some scattering for higher load levels, while condition G already shows scatter for load level $0.5\sigma_{ult}$. A clear tendency towards higher fatigue resistance for smoother surfaces is visible. The same is reported in surface fatigue studies that include post-processing [7,10,15,43,45,54]. A possible explanation for the scatter on E and G is that, for these sample groups, the non-linear low cycle regime is reached. On a Wöhler curve, the logarithmic-linear relationship between stress and number of cycles is only valid in the high cycle fatigue regime [62]. Another reason may be the coarser surface texture caused by lower line energy and the possibly uneven powder distribution, as previously mentioned in Sections 3.1 and 4.1.

The data in Figure 8 were previously published in [52], where the following sample naming was used: A—AsB-smooth, E—AsB-medium and G—AsB-rough. In [52], more detailed evaluations of surface texture and crack initiation are shown.

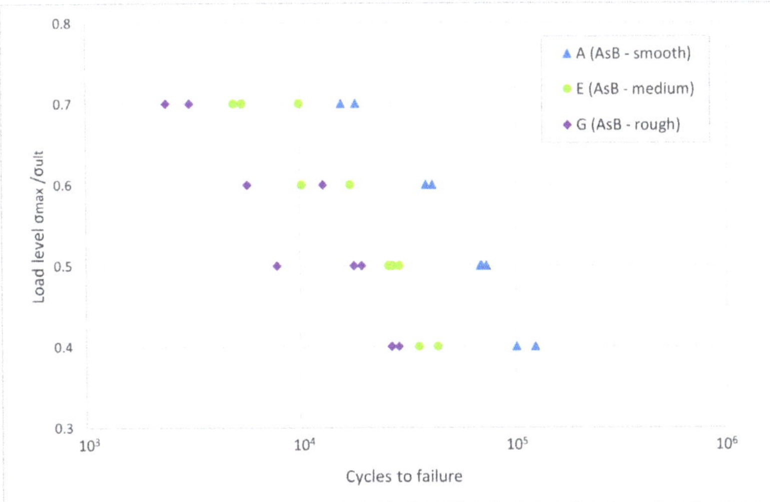

Figure 8. $\sigma - N$-curve for surface conditions A (AsB-smooth), E (AsB-medium) and G (AsB-rough), reference stress σ_{ult} = 392 MPa. Reproduced from [52].

4.2.1. Comparison with As-Built Surface Data from the Literature

To allow for comparison with surface quality data from the literature [39,46,51], Table 6 contains surface texture parameters generated with the respective cut-off values for one sample per surface condition.

Ra was determined from a 12 mm line profile, as indicated by Cacace et al. [46]. *Sa* and *Sv* were calculated from a 3 mm × 20 mm measured area. For measurement details, refer to [52]. A cut-off L-Filter of 0.8 mm was applied, as applied by Nasab et al. [51]. Please note that, deviating from Nasab et al., a least squares plane F-operation was used. The difference in F-operation is due to the sample geometries. This study assessed a flat sample geometry, while Nasab et al. evaluated cylindrical specimens.

Table 6. Surface texture parameters for comparison with the literature, cut-off 0.8 mm. Evaluated for one specimen per surface condition.

Surface Condition	*Ra*/μm	*Sa*/μm	*Sv*/μm
A (AsB-smooth)	3.153	3.478	20.09
E (AsB-medium)	5.649	6.987	93.78
G (AsB-rough)	7.362	9.316	96.57

Similarly to this paper, Nasab et al. [51] also used different as-built surfaces. However, taking a closer look at their considered surface conditions denoted S01, S05 and S07, they report larger *Sv*-values. Their best surface condition is S01 with $Sv = 112$ μm, having the order of magnitude of the roughest surface considered in this study, G (AsB-rough). S05 with $Sv = 190$ μm and S07 with $Sv = 205$ μm largely exceed the values presented in Table 6. In their work, they induced defects to demonstrate their influence on rotating bending fatigue. They suggested a minimum material removal based on surface texture parameter results to improve surface quality. In contrast, this work was aimed at producing high-quality surfaces (described by *Sa* and *Sk*) in as-built condition with no intention of surface post-processing.

Additionally, the samples investigated in this work have a smoother surface finish (see Table 6) in comparison with Cacace et al., who state an as-built $Ra > 10$ µm. They sandblasted the samples to achieve an $Ra < 10$ µm, as required for standard fatigue testing. The same holds for Grande et al., who reported an as-built Ra µm with $10 < Ra < 15$ µm [39,46].

4.2.2. Estimation of Stress at the Endurance Limit

The stress at the endurance limit was estimated based on the horizon method, using the logarithmic-linear equation

$$lg(N) = m \cdot lg(\sigma) + c \tag{1}$$

as described by Einbock [62]. For the sample groups A, E and G, the mean values for each load level were used to obtain their respective logarithmic-linear relations. The coefficients per surface condition are given in Table 7. For condition G, data points for load level 0.7 were in the LCF range with $N_{mean} = 2.7 \times 10^3$. They are most likely not on the linear part of the S–N curve and were therefore excluded from this calculation.

Table 7. Coefficients of logarithmic-linear equation $lg(N) = m \cdot lg(\sigma) + c$ for three surface conditions.

Coefficient	A (AsB-Smooth)	E (AsB-Medium)	G (AsB-Rough)
m	−3.292	−3.244	−2.892
c	3.794	3.364	3.283

From this equation, the stress at endurance limit σ_L was calculated for $N_{L1} = 10^6$, $N_{L2} = 2 \times 10^6$ and $N_{L3} = 10^7$ and is presented in Table 8 and Figure 9. N_{L1} to N_{L3} were chosen to allow for comparison with literature values [39,43,45,46,49,63].

$\sigma_{L2} = 49$ MPa for surface condition E corresponds well with the experimental findings of Denti and Sola [43] and Gatto et al. [45], who report mean values of 50 MPa for as-built specimens. Bassoli et al. [49] found a slightly higher σ_{L2} of 60 MPa, which is in between groups A and E. However, as they did not evaluate surface quality, no direct comparison is possible.

Cacace et al. and Grande et al. [39,46] found experimental endurance limit stress values at $N_{L3} = 10^7$ between 122 and 137 MPa for different positions on the build platform, being three times as high as found for A, the best performing condition studied here. As previously mentioned, they applied a sandblasting finish to meet the requirement for fatigue testing. Not only did this improve the surface finish, it also introduced compressive residual stresses, which prevent crack propagation [43,50].

Compared to the post-processed surface conditions presented by Denti and Sola [43] and Gatto et al. [45], the A condition's endurance stress matches the performance of laser shot processed and metal shot peened (S70) specimens.

Considering conventionally manufactured parts, Dezecot and Brochu estimated a fatigue strength of 73 MPa for as-cast AlSi7Mg0.6 material from investment casting [63] at $N_L = 10^6$. Surface condition A exceeds this value by 15%. This increased strength is supposedly related to the fine microstructure due to faster solidification of the material in the LPBF process.

Table 8. Estimated stress for different endurance limit values N_L .

Surface Condition	A (AsB-Smooth)	E (AsB-Medium)	G (AsB-Rough)
σ_{L1} at $N_{L1} = 10^6$/MPa	84	60	45
σ_{L2} at $N_{L2} = 2 \times 10^6$/MPa	68	49	35
σ_{L3} at $N_{L3} = 10^7$/MPa	42	30	20

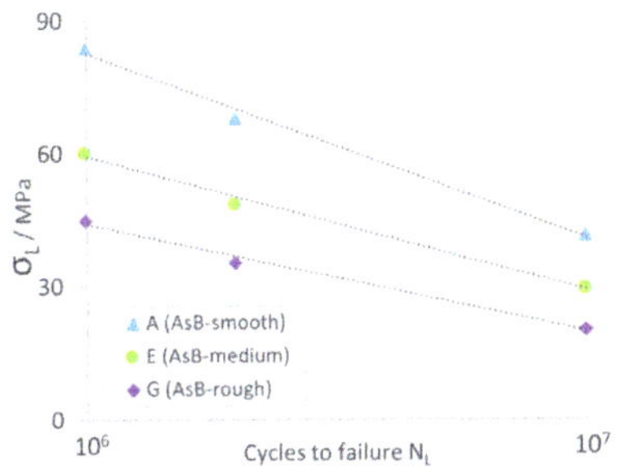

Figure 9. Estimated stress for different endurance limit values N_L.

4.2.3. Relationship of Surface Quality and Fatigue

In addition to the surface fatigue relationship shown in [52] for experimental values, this section presents data factorized to a load level of $0.5\sigma_{ult}$.

The factorization was done based on a linear regression across all data from the tested load levels. The exponential fit of *Svk* vs. the number of cycles to failure data in Figure 10 is described by

$$N = 159843e^{-0.273 \cdot Svk} \tag{2}$$

with $R^2 = 0.8721$. Numerical values for *Svk* are given in Table 9. Apart from the previously presented data for the A, E and G groups, there were also a few test results available from surface condition C samples, which were included in this fit.

The reduced valley depth *Svk*, derived from the material ratio curve, was chosen because it represents the valley population of a sample (within the measured area), as opposed to the common parameters for surface fatigue correlations *Sz* and *Sv*, which are individual extreme values and may not be representative of the considered surface.

This parameter choice is confirmed by the data shown in Tables 6 and 8. Surface conditions E and G have comparable *Sv* but different *Ra* and *Sa*. As previously discussed, the fatigue life for both groups clearly differs as well.

Furthermore, especially when looking at rougher surface conditions, cracks tend to start from multiple locations at the surface. Among the tested samples, this was observed for all group G specimens and half of group E specimens (for details, refer to [52]). Hence, considering the specific nature of typical LPBF-processed surfaces, it makes sense to consider more than just one extreme value per surface.

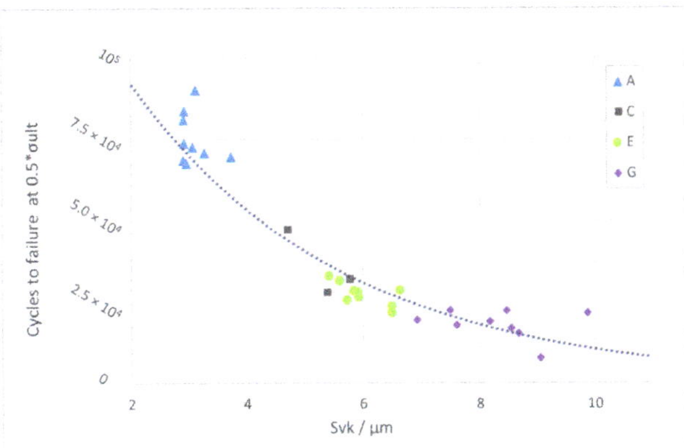

Figure 10. Exponential fit *Svk* vs. cycles to failure at $0.5\sigma_{ult}$.

Table 9. Svk for fatigue-tested samples.

Surface Condition	Mean Svk/µm	SD Svk/µm
A	3.105	0.271
C	5.296	0.539
E	6.028	0.431
G	8.321	0.890

5. Conclusions

This work aimed to produce samples with identical bulk and different surface quality, including high quality, to assess the impact of as-built surface texture on mechanical properties.

The evaluation of as-built surfaces was motivated by the desire to apply LPBF for complex geometries and inner surfaces, which may be complicated or infeasible to post-process. In addition, achieving the same surface finish and fatigue performance without post-processing saves time and resources.

The variation in surface texture was achieved by varying contour scan speed. The comparability of bulk quality for the different sample groups was confirmed by means of Archimedes' density and tensile testing. The endurance limit was estimated based on four tested fatigue load levels. The relationship between the reduced valley depth Svk for the different surface quality groups was shown using data factorized to load level $0.5\sigma_{ult}$.

The following main conclusions are derived from the presented work:

- Distinctive surface conditions with Sk (L-filter 0.25 mm) between 4 µm and 16 µm were produced.
- All tested specimens have a density > 99%; thus, the influence of contour scan parameters is considered insignificant regarding density.
- The ultimate tensile strength of 393 ± 9.98 MPa was found to be independent of contour scan variation.
- Optimized contour scan parameters result in as-built quality superior to some post-processed surfaces, enabling the reduction of processing steps and time.
- Condition A reaches a fatigue resistance of 84 MPa at 10^6 cycles, exceeding values for as-cast and some surface post-processed literature results.
- The reduced valley depth Svk results in a good fit across the groups for the factorized surface fatigue relation. Therefore, Svk was found to be a suitable parameter to describe surface quality.

Author Contributions: Conceptualization, T.B.; methodology, T.B., M.A. and H.B.; formal analysis, T.B.; investigation, T.B.; resources, M.A., H.B. and B.M.; data curation, T.B.; writing—original draft preparation, T.B.; writing—review and editing, T.B., M.A., H.B. and B.M.; visualization, T.B.; funding acquisition, M.A., H.B. and B.M. All authors have read and agreed to the published version of the manuscript.

Funding: This research received no external funding.

Data Availability Statement: Data are available from the corresponding author upon request.

Acknowledgments: The authors thank Inga Meyenborg for manufacturing the samples and performing density measurements and Annette Schwingen for performing tensile testing.

Conflicts of Interest: The authors declare no conflict of interest.

Appendix A

Table A1. *Sa* and *Sk* for cuboid samples, 10 surface conditions L-filter 0.25 mm, S-filter 20 μm. With pre-sinter (right), without pre-sinter (left).

Parameter Set Scan Speed/mm/s	J 1800	H 1200	F 900	D 600	B 300	A 300	C 600	E 900	G 1200	I 1800
Sa/μm, Mean (N = 3)	5.94	3.88	2.33	1.82	1.54	1.42	1.63	2.17	3.61	5.85
Sa/μm, SD	0.10	0.34	0.13	0.08	0.06	0.08	0.04	0.07	0.15	0.61
Sk/μm, Mean (N = 3)	15.93	10.25	6.66	5.35	4.53	4.30	4.88	6.49	10.50	16.09
Sk/μm, SD	0.26	0.67	0.24	0.15	0.05	0.18	0.15	0.06	0.37	1.18

Table A2. Ultimate tensile strength and relative density of tensile samples. Reference density 2.68 g/cm^3. Seven surface conditions.

Parameter Set N	A 6	C 6	D 6	E 6	G 6	H 5	J 5
σ_{ult}/MPa, Mean	391.63	405.90	401.42	400.87	388.79	387.96	374.21
σ_{ult}/MPa, SD	5.26	10.33	7.37	14.54	11.04	15.55	16.33
Relative Density/%, Mean	99.85	99.68	99.52	99.61	99.69	99.58	99.74
Relative Density/%, SD	0.17	0.15	0.18	0.11	0.17	0.28	0.20

References

1. Uriondo, A.; Esperon-Miguez, M.; Perinpanayagam, S. The present and future of additive manufacturing in the aerospace sector: A review of important aspects. *Proc. Inst. Mech. Eng. Part G J. Aerosp. Eng.* **2015**, *229*, 2132–2147. [CrossRef]
2. Herzog, D.; Seyda, V.; Wycisk, E.; Emmelmann, C. Additive manufacturing of metals. *Acta Mater.* **2016**, *117*, 371–392. [CrossRef]
3. Manfredi, D.; Bidulský, R. Laser powder bed fusion of aluminum alloys. *Acta Metall. Slovaca* **2017**, *23*, 276–282. [CrossRef]
4. Aboulkhair, N.T.; Simonelli, M.; Parry, L.; Ashcroft, I.; Tuck, C.; Hague, R. 3D printing of Aluminium alloys: Additive Manufacturing of Aluminium alloys using selective laser melting. *Prog. Mater. Sci.* **2019**, *106*, 100578. [CrossRef]
5. O'Brien, M. Existing standards as the framework to qualify additive manufacturing of metals. In Proceedings of the 2018 IEEE Aerospace Conference, Big Sky, MT, USA, 3–10 March 2018, pp. 1–10.
6. Mohd Yusuf, S.; Cutler, S.; Gao, N. The impact of metal additive manufacturing on the aerospace industry. *Metals* **2019**, *9*, 1286. [CrossRef]
7. Bagehorn, S.; Wehr, J.; Maier, H. Application of mechanical surface finishing processes for roughness reduction and fatigue improvement of additively manufactured Ti-6Al-4V parts. *Int. J. Fatigue* **2017**, *102*, 135–142. [CrossRef]
8. Nicoletto, G.; Konečná, R.; Frkáň, M.; Riva, E. Surface roughness and directional fatigue behavior of as-built EBM and DMLS Ti6Al4V. *Int. J. Fatigue* **2018**, *116*, 140–148. [CrossRef]
9. Frkan, M.; Konecna, R.; Nicoletto, G.; Kunz, L. Microstructure and fatigue performance of SLM-fabricated Ti6Al4V alloy after different stress-relief heat treatments. *Transp. Res. Procedia* **2019**, *40*, 24–29. [CrossRef]
10. Kahlin, M.; Ansell, H.; Basu, D.; Kerwin, A.; Newton, L.; Smith, B.; Moverare, J. Improved fatigue strength of additively manufactured Ti6Al4V by surface post processing. *Int. J. Fatigue* **2020**, *134*, 105497. [CrossRef]
11. Cersullo, N.; Mardaras, J.; Emile, P.; Nickel, K.; Holzinger, V.; Hühne, C. Effect of Internal Defects on the Fatigue Behavior of Additive Manufactured Metal Components: A Comparison between Ti6Al4V and Inconel 718. *Materials* **2022**, *15*, 6882. [CrossRef]
12. Gockel, J.; Sheridan, L.; Koerper, B.; Whip, B. The influence of additive manufacturing processing parameters on surface roughness and fatigue life. *Int. J. Fatigue* **2019**, *124*, 380–388. [CrossRef]
13. Nicoletto, G. Smooth and notch fatigue behavior of selectively laser melted Inconel 718 with as-built surfaces. *Int. J. Fatigue* **2019**, *128*, 105211. [CrossRef]
14. Uriati, F.; Nicoletto, G. A comparison of Inconel 718 obtained with three L-PBF production systems in terms of process parameters, as-built surface quality, and fatigue performance. *Int. J. Fatigue* **2022**, *162*, 107004. [CrossRef]
15. Afkhami, S.; Dabiri, M.; Piili, H.; Björk, T. Effects of manufacturing parameters and mechanical post-processing on stainless steel 316L processed by laser powder bed fusion. *Mater. Sci. Eng. A* **2021**, *802*, 140660. [CrossRef]
16. Obeidi, M.A.; Mhurchadha, S.M.U.; Raghavendra, R.; Conway, A.; Souto, C.; Tormey, D.; Ahad, I.U.; Brabazon, D. Comparison of the porosity and mechanical performance of 316L stainless steel manufactured on different laser powder bed fusion metal additive manufacturing machines. *J. Mater. Res. Technol.* **2021**, *13*, 2361–2374. [CrossRef]
17. Alaimo, G.; Carraturo, M.; Korshunova, N.; Kollmannsberger, S. Numerical evaluation of high cycle fatigue life for additively manufactured stainless steel 316L lattice structures: Preliminary considerations. *Mater. Des. Process. Commun.* **2021**, *3*, e249. [CrossRef]

18. Korshunova, N.; Alaimo, G.; Hosseini, S.B.; Carraturo, M.; Reali, A.; Niiranen, J.; Auricchio, F.; Rank, E.; Kollmannsberger, S. Image-based numerical characterization and experimental validation of tensile behavior of octet-truss lattice structures. *Addit. Manuf.* **2021**, *41*, 101949. [CrossRef]
19. Magarò, P.; Alaimo, G.; Carraturo, M.; Sgambitterra, E.; Maletta, C. A novel methodology for the prediction of the stress–strain response of laser powder bed fusion lattice structure based on a multi-scale approach. *Mater. Sci. Eng. A* **2023**, *863*, 144526. [CrossRef]
20. Sercombe, T.B.; Li, X. Selective laser melting of aluminium and aluminium metal matrix composites. *Mater. Technol.* **2016**, *31*, 77–85. [CrossRef]
21. Aboulkhair, N.T.; Everitt, N.M.; Maskery, I.; Ashcroft, I.; Tuck, C. Selective laser melting of aluminum alloys. *MRS Bull.* **2017**, *42*, 311–319. [CrossRef]
22. DebRoy, T.; Wei, H.; Zuback, J.; Mukherjee, T.; Elmer, J.; Milewski, J.; Beese, A.M.; Wilson-Heid, A.d.; De, A.; Zhang, W. Additive manufacturing of metallic components–process, structure and properties. *Prog. Mater. Sci.* **2018**, *92*, 112–224. [CrossRef]
23. Kusoglu, I.M.; Gökce, B.; Barcikowski, S. Research trends in laser powder bed fusion of Al alloys within the last decade. *Addit. Manuf.* **2020**, *36*, 101489. [CrossRef]
24. Rometsch, P.A.; Zhu, Y.; Wu, X.; Huang, A. Review of high-strength aluminium alloys for additive manufacturing by laser powder bed fusion. *Mater. Des.* **2022**, *219*, 110779. [CrossRef]
25. Kotadia, H.; Gibbons, G.; Das, A.; Howes, P. A review of Laser Powder Bed Fusion Additive Manufacturing of aluminium alloys: Microstructure and properties. *Addit. Manuf.* **2021**, *46*, 102155. [CrossRef]
26. Lathabai, S. Additive manufacturing of aluminium-based alloys and composites. *Fundam. Alum. Metall.* **2018**, 47–92. [CrossRef]
27. Prashanth, K.G.; Scudino, S.; Klauss, H.J.; Surreddi, K.B.; Löber, L.; Wang, Z.; Chaubey, A.K.; Kühn, U.; Eckert, J. Microstructure and mechanical properties of Al–12Si produced by selective laser melting: Effect of heat treatment. *Mater. Sci. Eng. A* **2014**, *590*, 153–160. [CrossRef]
28. Kimura, T.; Nakamoto, T. Microstructures and mechanical properties of A356 (AlSi7Mg0. 3) aluminum alloy fabricated by selective laser melting. *Mater. Des.* **2016**, *89*, 1294–1301. [CrossRef]
29. Chen, B.; Moon, S.; Yao, X.; Bi, G.; Shen, J.; Umeda, J.; Kondoh, K. Strength and strain hardening of a selective laser melted AlSi10Mg alloy. *Scr. Mater.* **2017**, *141*, 45–49. [CrossRef]
30. Rao, J.H.; Zhang, Y.; Fang, X.; Chen, Y.; Wu, X.; Davies, C.H. The origins for tensile properties of selective laser melted aluminium alloy A357. *Addit. Manuf.* **2017**, *17*, 113–122. [CrossRef]
31. Yang, K.V.; Rometsch, P.; Davies, C.; Huang, A.; Wu, X. Effect of heat treatment on the microstructure and anisotropy in mechanical properties of A357 alloy produced by selective laser melting. *Mater. Des.* **2018**, *154*, 275–290. [CrossRef]
32. Takata, N.; Liu, M.; Kodaira, H.; Suzuki, A.; Kobashi, M. Anomalous strengthening by supersaturated solid solutions of selectively laser melted Al–Si-based alloys. *Addit. Manuf.* **2020**, *33*, 101152. [CrossRef]
33. Pereira, J.C.; Gil, E.; Solaberrieta, L.; San Sebastián, M.; Bilbao, Y.; Rodríguez, P.P. Comparison of AlSi7Mg0.6 alloy obtained by selective laser melting and investment casting processes: Microstructure and mechanical properties in as-built/as-cast and heat-treated conditions. *Mater. Sci. Eng. A* **2020**, *778*, 139124. [CrossRef]
34. Salandari-Rabori, A.; Wang, P.; Dong, Q.; Fallah, V. Enhancing as-built microstructural integrity and tensile properties in laser powder bed fusion of AlSi10Mg alloy using a comprehensive parameter optimization procedure. *Mater. Sci. Eng. A* **2021**, *805*, 140620. [CrossRef]
35. Liu, M.; Takata, N.; Suzuki, A.; Kobashi, M.; Kato, M. Enhancement in strength and ductility of laser powder bed fused Al–12Si alloy by introducing nanoscale precipitates. *Addit. Manuf. Lett.* **2021**, *1*, 100008. [CrossRef]
36. Zhang, B.; Wei, W.; Shi, W.; Guo, Y.; Wen, S.; Wu, X.; Gao, K.; Rong, L.; Huang, H.; Nie, Z. Effect of heat treatment on the microstructure and mechanical properties of Er-containing Al–7Si–0.6 Mg alloy by laser powder bed fusion. *J. Mater. Res. Technol.* **2022**, *18*, 3073–3084. [CrossRef]
37. Uzan, N.E.; Shneck, R.; Yeheskel, O.; Frage, N. Fatigue of AlSi10Mg specimens fabricated by additive manufacturing selective laser melting (AM-SLM). *Mater. Sci. Eng. A* **2017**, *704*, 229–237. [CrossRef]
38. Zhang, C.; Zhu, H.; Liao, H.; Cheng, Y.; Hu, Z.; Zeng, X. Effect of heat treatments on fatigue property of selective laser melting AlSi10Mg. *Int. J. Fatigue* **2018**, *116*, 513–522. [CrossRef]
39. Grande, A.; Cacace, S.; Demir, A.G.; Sala, G. Fracture and fatigue behaviour of AlSi7Mg0.6 produced by Selective Laser Melting: Effects of thermal-treatments. In Proceedings of the 25th International Congress, Rome, Italy, 9–12 September 2019.
40. Mauduit, A.; Gransac, H.; Auguste, P.; Pillot, S. Study of AlSi7Mg0.6 Alloy by Selective Laser Melting: Mechanical Properties, Microstructure, Heat Treatment. *J. Cast. Mater. Eng.* **2019**, *3*, 1. [CrossRef]
41. Han, Q.; Jiao, Y. Effect of heat treatment and laser surface remelting on AlSi10Mg alloy fabricated by selective laser melting. *Int. J. Adv. Manuf. Technol.* **2019**, *102*, 3315–3324. [CrossRef]
42. Aboulkhair, N.T.; Maskery, I.; Tuck, C.; Ashcroft, I.; Everitt, N.M. Improving the fatigue behaviour of a selectively laser melted aluminium alloy: Influence of heat treatment and surface quality. *Mater. Des.* **2016**, *104*, 174–182. [CrossRef]
43. Denti, L.; Sola, A. On the effectiveness of different surface finishing techniques on A357. 0 parts produced by laser-based powder bed fusion: Surface roughness and fatigue strength. *Metals* **2019**, *9*, 1284. [CrossRef]
44. Hamidi Nasab, M.; Giussani, A.; Gastaldi, D.; Tirelli, V.; Vedani, M. Effect of surface and subsurface defects on fatigue behavior of AlSi10Mg alloy processed by laser powder bed fusion (L-PBF). *Metals* **2019**, *9*, 1063. [CrossRef]

45. Gatto, A.; Bassoli, E.; Denti, L.; Sola, A.; Tognoli, E.; Comin, A.; Porro, J.A.; Cordovilla, F.; Angulo, I.; Ocaña, J.L. Effect of Three Different Finishing Processes on the Surface Morphology and Fatigue Life of A357. 0 Parts Produced by Laser-Based Powder Bed Fusion. *Adv. Eng. Mater.* **2019**, *21*, 1801357. [CrossRef]
46. Cacace, S.; Demir, A.G.; Sala, G.; Grande, A.M. Influence of production batch related parameters on static and fatigue resistance of LPBF produced AlSi7Mg0. 6. *Int. J. Fatigue* **2022**, *165*, 107227. [CrossRef]
47. de Menezes, J.T.O.; Castrodeza, E.M.; Casati, R. Effect of build orientation on fracture and tensile behavior of A357 Al alloy processed by Selective Laser Melting. *Mater. Sci. Eng. A* **2019**, *766*, 138392. [CrossRef]
48. Denti, L. Additive manufactured A357. 0 samples using the laser powder bed fusion technique: Shear and tensile performance. *Metals* **2018**, *8*, 670. [CrossRef]
49. Bassoli, E.; Denti, L.; Comin, A.; Sola, A.; Tognoli, E. Fatigue Behavior of As-Built L-PBF A357.0 Parts. *Metals* **2018**, *8*, 634. [CrossRef]
50. Mower, T.M.; Long, M.J. Mechanical behavior of additive manufactured, powder-bed laser-fused materials. *Mater. Sci. Eng. A* **2016**, *651*, 198–213. [CrossRef]
51. Nasab, M.H.; Romano, S.; Gastaldi, D.; Beretta, S.; Vedani, M. Combined effect of surface anomalies and volumetric defects on fatigue assessment of AlSi7Mg fabricated via laser powder bed fusion. *Addit. Manuf.* **2020**, *34*, 100918.
52. Buchenau, T.; Amkreutz, M.; Brüning, H. Surface texture and high cycle fatigue of as-built metal additive AlSi7Mg0. 6. *J. Addit. Manuf. Technol.* **2021**, *1*, 531–531. [CrossRef]
53. Greitemeier, D.; Dalle Donne, C.; Syassen, F.; Eufinger, J.; Melz, T. Effect of surface roughness on fatigue performance of additive manufactured Ti–6Al–4V. *Mater. Sci. Technol.* **2016**, *32*, 629–634. [CrossRef]
54. Raab, M.; Bambach, M. Fatigue properties of Scalmalloy® processed by laser powder bed fusion in as-built, chemically and conventionally machined surface condition. *J. Mater. Process. Technol.* **2023**, *311*, 117811. [CrossRef]
55. Buchenau, T.; Cersullo, N.; Mardaras, J.; Emile, P.; Lafue, V.; Brüning, H.; Amkreutz, M.; Hühne, C. Fatigue and surface texture of post-processed metal additive Ti-6Al-4V. In Proceedings of the 17th European Conference on Spacecraft Structures, Materials and Environmental Testing, Toulouse, France, 28–30 March 2023.
56. *ASTM E466-15*; Standard Practice for Conducting Force Controlled Constant Amplitude Axial Fatigue Tests of Metallic Materials. ASTM: West Conchhocken, PA, USA, 2015. . [CrossRef]
57. Townsend, A.; Senin, N.; Blunt, L.; Leach, R.; Taylor, J. Surface texture metrology for metal additive manufacturing: A review. *Precis. Eng.* **2016**, *46*, 34–47. [CrossRef]
58. Steinhilper, W.; Sauer, B.; Feldhusen, J. *Konstruktionselemente des Maschinenbaus 1: Grundlagen der Berechnung und Gestaltung von Maschinenelementen*; Springer: Berlin/Heidelberg, Germany, 2008.
59. *ISO 25178*; Geometric Product Specifications (GPS)—Surface Texture: Areal—Part 2: Terms, Definitions and Surface Texture Parameters. ISO: Geneva, Switzerland, 2022.
60. *ISO 3369:2006*; Impermeable Sintered Metal Materials and Hardmetals—Determination of Density. ISO: Geneva, Switzerland, 2006.
61. *ASTM E8M*; Standard Test Methods for Tension Testing of Metallic Materials. ASTM: West Conchhocken, PA, USA, 2016. [CrossRef]
62. Einbock, S. *Statistics of Metal Fatigue in Engineering: Planning and Analysis of Metal Fatigue Tests*; BoD–Books on Demand: Norderstedt, Germany, 2018.
63. Dezecot, S.; Brochu, M. Microstructural characterization and high cycle fatigue behavior of investment cast A357 aluminum alloy. *Int. J. Fatigue* **2015**, *77*, 154–159. [CrossRef]

materials

MDPI

Article

Effect of Fractal Ceramic Structure on Mechanical Properties of Alumina Ceramic–Aluminum Composites

Xianjun Zeng [1], Qiang Jing [1], Jianwei Sun [1] and Jinyong Zhang [1,2,*]

1 State Key Laboratory of Advanced Technology for Materials Synthesis and Processing, Wuhan University of Technology, Wuhan 430070, China; 13657225706@163.com (X.Z.)
2 Hubei Longzhong Laboratory, Xiangyang 441000, China
* Correspondence: jyzhang@whut.edu.cn

Abstract: In conventional ceramic–metal matrix composites, with the addition of the ceramic phase, although it can significantly improve the performance of the material in one aspect, it tends to weaken some of the excellent properties of the metal matrix as well. In order to meet the high toughness and high strength requirements of composites for practical production applications, researchers have searched for possible reinforcing structures from nature. They found that fractal structures, which are widely found in nature, have the potential to improve the mechanical properties of materials. However, it is often not feasible to manufacture these geometric structures using conventional processes. In this study, alumina ceramic fractal structures were prepared by 3D printing technology, and aluminum composites containing fractal ceramic structures were fabricated by spark plasma sintering technology. We have studied the effect of the fractal structure of alumina ceramics on the mechanical properties of composites. The compression strength of samples was measured by a universal testing machine and the torsional properties of samples were measured by a torsional modulus meter. The results show that a fractal structure improves the compressive strength of aluminum/alumina ceramic composites by 10.97% and the torsional properties by 17.45%. The results of the study will provide a new method for improving the mechanical properties of materials.

Keywords: 3D printing; fractal structure; mechanical property

Citation: Zeng, X.; Jing, Q.; Sun, J.; Zhang, J. Effect of Fractal Ceramic Structure on Mechanical Properties of Alumina Ceramic–Aluminum Composites. *Materials* **2023**, *16*, 2296. https://doi.org/10.3390/ma16062296

Academic Editor: Alberto Ortona

Received: 23 February 2023
Revised: 7 March 2023
Accepted: 9 March 2023
Published: 13 March 2023

1. Introduction

Ceramic-reinforced metal matrix composites integrate many excellent characteristics of metal and ceramic materials. While maintaining the good plastic toughness of metal materials, they have the high hardness, high strength, good wear resistance, and corrosion resistance of ceramic materials [1–3]. Ceramic-reinforced aluminum matrix composites are the most representative material of ceramic-reinforced metal matrix composites. A ceramic-reinforced aluminum matrix composite has the characteristics of high specific strength, specific stiffness, elastic modulus, wear resistance, and good dimensional stability [4]. It has great practicability and a broad application prospect in aerospace, automobile manufacturing, precision instruments, electronic packaging, sports equipment, and other fields [5]. However, in practical studies, many researchers have found that with the addition of the ceramic phase, the overall hardness and strength of the composite material has increased, while some of the excellent properties of the original metal matrix, such as toughness and plasticity, have decreased [6]. In order to meet the requirements of high strength, high toughness, high hardness, and excellent torsional resistance of composite materials in practical applications, we need to consider the structure of ceramic materials and composite material fabrication. By observing a variety of phenomena in nature, researchers have found some patterns (Figure 1). For example, mature trees are often able to resist the strong torsional forces caused by strong winds, the threaded structure of shells can make them significantly more resistant to compression, and honeycombs can have huge space inside while still having some mechanical strength [7–9]. Through further study of these examples, it is believed that some special and ordered structures in trees and shells are the main reason why they have such excellent mechanical properties. This special and ordered structure has been referred to as the fractal structure (Figure 1).

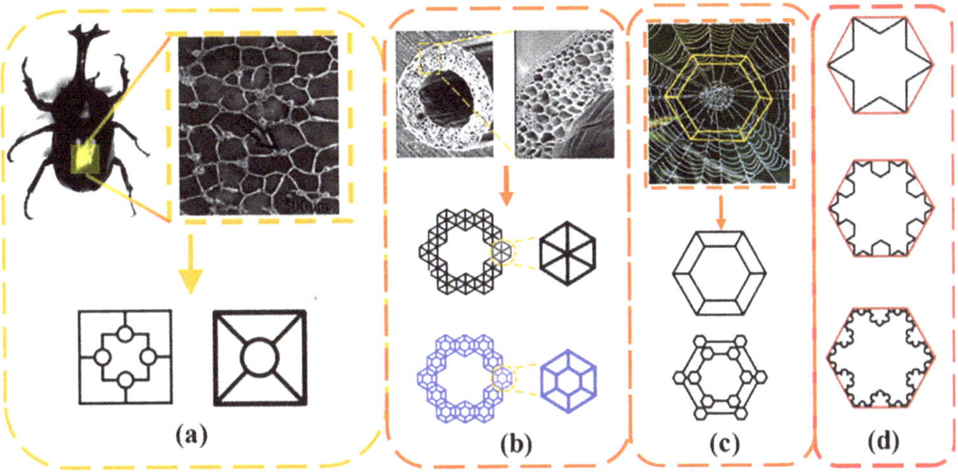

Figure 1. Bio-inspired structures and fractal structures. (**a**) Insect-inspired fractal structure, (**b**) the presence of fractal structures in wood, (**c**) spider's web of fractal structure, (**d**) the evolution of fractal structures.

Fractal is usually defined as "a rough or fragmented geometric shape that can be divided into several parts, each of which is a reduced shape of the whole", i.e., it has the property of self-similarity. The term fractal was originally used by mathematician Benoit Mandelbrot in 1975, to denote a series of objects with specific characteristics, such as self-similar structure and shape geometry, on all magnification scales [10]. In the past decade, people's interest in using fractal structures to explore the improvement of mechanical and mechanical properties has greatly increased. According to the definition of fractal structures, we know that self-similarity and iterative generation are two important features of fractal structures. Examples of the results of studies on the effects of fractal structure include, the interlocking properties of hierarchical fractal structures providing better load distribution and energy absorption. Farina et al. [11] analyzed the bending behavior of cementitious composites reinforced by straight fractal titanium alloy rods, and the interlocking mechanism between the fractal rods and the matrix resulted in a 152% increase in bending strength. Another example of the application of fractal structures is the optimization of fluid distribution in tree channels. Wang et al. [12] showed that applying fractal structures to heat exchanger channels can improve the heat transfer coefficient and reduce the pressure drop, compared to conventional heat exchangers. Research on the impact resistance of tree-shaped fractal structures is more in-depth. San Ha Ngoc et al. [13] studied the kinetic energy absorption capacity of thin-walled tubes with bionic tree-like cross-sections. Other projects have developed impact-resistant protection devices based on fractal structures, such as fractal honeycomb, Koch curve, side fractal shapes, and Sierpinski shapes [14,15].

Although these properties have been demonstrated, current applications are limited by the fact that these structures, especially in three-dimensional geometries, are difficult to produce, or often impossible to achieve, with conventional manufacturing techniques. Over the past few decades, the development of 3D printing processes has opened up a number of design opportunities that offer more possibilities for manufacturing such complex structures. Stereolithography appearance (SLA) is one of the most widely used 3D printing techniques [16]. Marco Viccica et al. used SLA technology to fabricate a 3D cross-base fractal structure for shock absorption, and studied the mechanical properties of a 3D Greek cross fractal [17]. They conducted numerical studies on the mechanical behavior of the structure under quasi-static and dynamic compression loads, established a material model, and verified the correctness of the model through experiments. It was found that the energy absorption effect of the three-dimensional cross-base fractal structure was 77% higher than that of traditional foam. Wu et al. [18] proposed a new energy-absorbing protective structure, developed by using SLA technology. They established triangular, square, and pentagonal tree fractal structures, and studied their mechanical behavior and deformation process by quasi-static axial fracture tests. These showed that, compared with a single-wall structure, the tree-shaped fractal structure has greater potential to improve the specific energy absorption and resist impact forces. The tree-shaped fractal design promotes the deformation stability of the thin-wall structure. Compared with the single-wall structure, the energy absorption efficiency and load stability of the tree-shaped structure are improved. The tree fractal design improves the deformation stability of single-walled structures by controlling the geometry and material distribution of the structures.

In order to better explore the effect of fractal structure on the mechanical properties of materials, four kinds of alumina ceramic/aluminum composites, containing a fractal structure, were designed and fabricated. The compression strength of four kinds of samples was measured by a universal testing machine. The torsional properties of the different samples were measured by using a torsional modulus meter. The influence mechanism of the fractal structure was also explored and analyzed.

2. Structure Description

Sierpinski Fractal Structure

The Sierpinski polygon, a self-similar structure discovered by Waclaw Sierpinski in the early 20th century, is one of the most commonly used fractal patterns [19]. However, the mechanical properties of this Sierpinski fractal pattern have not been considered, which has prompted the present study to propose advanced shapes, combined with a Sierpinski fractal, to make composite materials.

Triangle Sierpinski fractals and square Sierpinski fractals are two classical fractals in the category of Sierpinski fractals. Their formation processes are shown in Figure 2a,b, respectively.

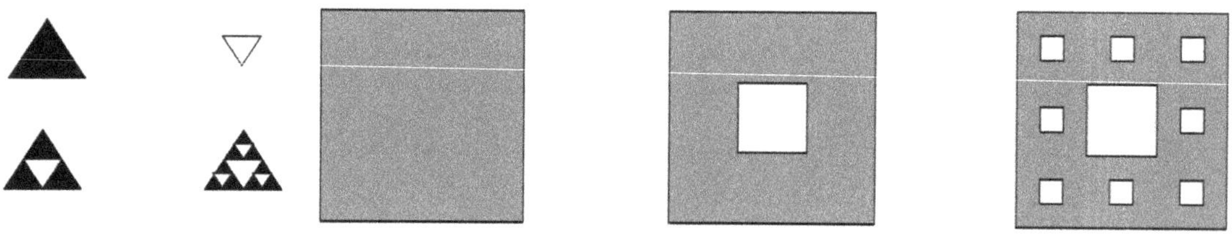

(a) (b)

Figure 2. (**a**) Triangle Sierpinski fractal, (**b**) square Sierpinski fractal.

Among them, the formation process of the triangle Sierpinski fractal is as follows: the initial structure is an equilateral triangle. In the first step, the initial structure is divided into four identical smaller equilateral triangles, and in the second step, the middle is removed, and a first-order Sierpinski fractal structure is generated, as shown in Figure 2a. For the remaining three small equilateral triangles in the above figure, the above two steps are repeated. The results of this, as shown in Figure 2a, are second-order Sierpinski fractals. The fractal structure of other orders can be obtained in accordance with the above two steps.

The formation process of a quadrilateral Sierpinski fractal is as follows: the initial figure is a square structure. In the first step, the initial square is divided into nine identical smaller squares, and in the second step, a small square in the middle is removed, to generate a first-order Sierpinski fractal structure, as shown in Figure 2b. For the remaining eight small squares, the structure obtained by repeating the operation described above is shown in Figure 2b. Other structures of different orders can be obtained by repeating this process.

In this study, we designed four Sierpinski fractals: triangle (Sier 3), square (Sier 4), pentagon (Sier 5), and hexagon (Sier 6). The structure of each polygon follows the fractal law of Sierpinski. Figure 3 shows the schematic diagram of the four fractal design models. The radius of the outer circle of the largest polygon is 50 mm.

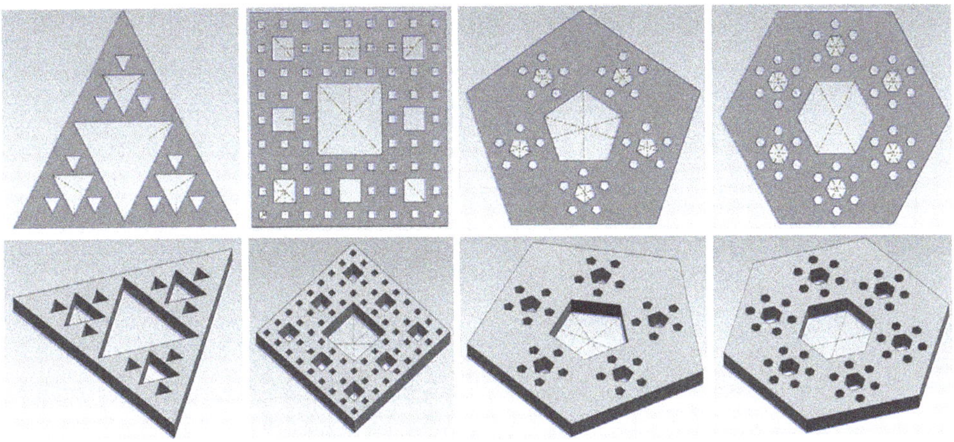

Figure 3. Sierpinski fractal structure model diagram, from left to right, triangle, square, pentagon, and hexagon.

3. Experimental Procedure

Regarding the design preparation and testing of alumina ceramic/aluminum composites, the experiments performed can be broadly divided into the following steps.

3.1. 3D Printing Experiment

(1) A fractal structure was designed, using the Solidwork modeling software (Siemens AG, GER), and imported into a 3D printer.

(2) The CeraBuilder160Pro ceramic laser 3D printer (Hubei Wuhan ILaser Inc., China) and ceramic paste (iLaser Inc., CHN) were used, to fabricate the fractal structure alumina ceramics. The specific printing parameters were as follows: the thickness of the printing layer was 0.1 mm, the size of the laser spot was 140 μm, and the working temperature was 25 °C. Then, the 3D printed-Al_2O_3 ceramics were put into an air atmosphere box-type furnace (ksl1700x, Hefei Kejing Materials Technology Co., Ltd., China) to remove photosensitive resin. The steps to remove the photosensitive resin were: first, the temperature was heated from room temperature to 300 °C at a rate of 1 °C/min and then kept stable for 120 min. Second, the temperature was heated from 300 °C to 550 °C at a rate of 0.5 °C/min and then kept stable for 120 min. Finally, the temperature was heated from 550 °C to 800 °C at a rate of 2 °C/min, then kept it at 800 °C for 90 min, and then cooled to room temperature naturally.

(3) After removal of the photosensitive resin, the sample was sintered in a vacuum sintering furnace, first at 3 °C/min to 1250 °C, held for 60 min, then at 2 °C/min to 1600 °C, held for 90 min, then cooled at 2 °C/min. After cooling to 300 °C in the furnace, the fractal Al_2O_3 ceramic structure was fabricated (Figure 4a). The entire construction process is shown in Figure 5.

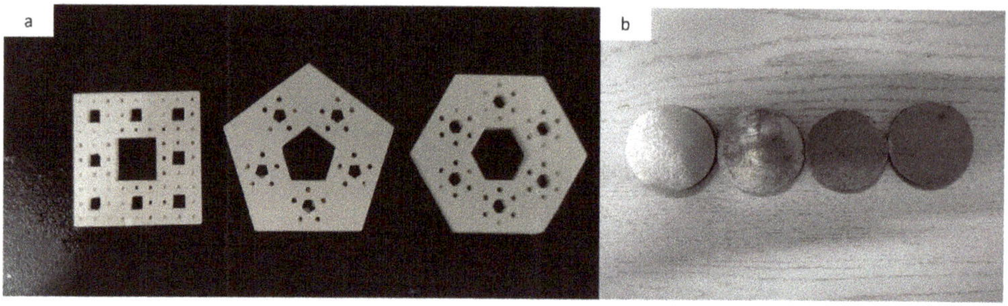

Figure 4. (**a**)The appearance of fractal structure (**b**)composite disk made with SPS.

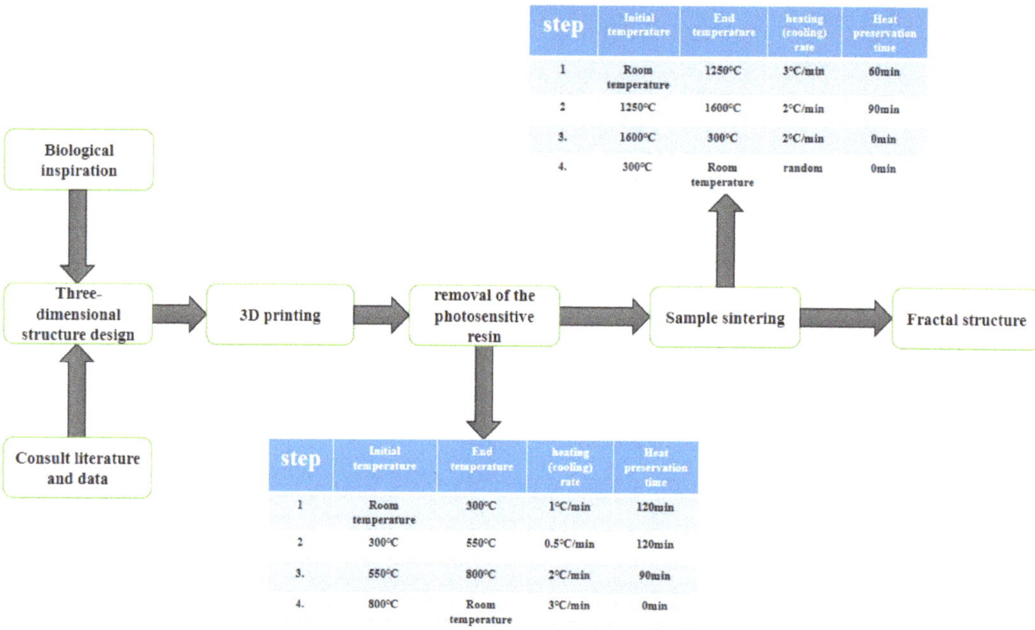

Figure 5. The process of making the fractal structure.

3.2. Fabrication of Alumina Ceramic/Aluminum Composites

Powder metallurgy (PM) is a flexible technology for manufacturing near-clean shape products. Traditional powder metallurgy usually involves three main steps: mixing of the metal powder with a reinforcing agent, compaction, and sintering at high temperature [20]. However, due to the absence of local heating, conventional PM products have high porosity, which further reduces the performance of these products [21]. Spark plasma sintering (SPS) is an effective method for preparing composite materials [22]. Short time densification and high local temperature limit undesired grain growth during SPS [23]. In addition, the alumina layer on the initial particles is removed by spark application, which significantly eliminates the porosity of the manufactured product [24]. Therefore, in this study, we chose to use SPS to make the alumina ceramic/aluminum composite materials.

Commercial spherical aluminum powder (Shanghai Hushi, purity ~99.7%, particle size ~38 μm) was used as the starting material. Table 1 showed the parameters of the raw materials used in this experiment. Table 2 shows the chemical composition of the aluminum powder. The experiment was divided into the following steps. First, 25 g pure aluminum powder and alumina ceramic structure were charged and compacted in an SPS graphite mold, at room temperature and pressure, and then sintered in an SPS sintering furnace. Sintering was carried out in a vacuum atmosphere, with a heating rate of 50 °C/min and a pressure of 45 MPa. Finally, the sintering temperature reached 550 °C. After reaching 550 °C, the sintering temperature was held constant for 10 min. When the insulation was completed, the sample was cooled to room temperature, under a pressure of 45 MPa, for 30 min. After the above steps, a disc-shaped sample, with a diameter of 50 mm and a thickness of about 6 mm, was produced (Figure 4).

Table 1. The parameters of the raw materials used in this experiment.

Powder Type	Particle Size/µm	Purity/%	Manufacturer
Al	38-40	99.7	Shanghai Hushi
Al_2O_3	2~3	99.9	Sumitomo chemical company

Table 2. Chemical composition of the pure aluminum powder used. (Information provided by supplier).

Al	Si	Fe	Cu	Zn	Mn	Mg	Ni
>99.7%	<0.15%	<0.2%	<0.02%	<0.02%	<0.02%	<0.02%	<0.02%

In order to compare the performance difference between an aluminum matrix composite made by adding a fractal structure and traditional methods, we fabricated composites with the same size and weight under the same sintering system. First, we used a balance to measure the weight of each fractal structure, and then weighed the alumina ceramic powder equal to the weight of the fractal structure. Secondly, we mixed alumina ceramic powder and 25 g aluminum powder evenly, by the ball milling method. The uniform powder was charged and compacted in an SPS graphite mold, at room temperature and pressure, and then sintered in an SPS sintering furnace. The sintering system was the same as above.

3.3. *Test Means*

Using an MSA324S-000-DU balance, the density of each sample, containing a different fractal skeleton, was measured by the Archimedes drainage method. The thickness of each sample before and after sintering was measured by Vernier caliper. In order to observe the microstructure of the aluminum matrix composite with a fractal structure, the middle part of the sintered sample was selected and cut into strips with dimensions of 3 mm × 4.5 mm × 36 mm. The cross-section of the sintered aluminum matrix composite was ground with sandpaper, and polished with diamond suspension after grinding. A scanning electron microscope (SEM, Hitachi 3400, Japan) was used to observe the fracture and surface morphology of each sample. The compression properties of all samples were tested on an electronic universal testing machine (MTS810, MTS, USA, loading rate 0.5 mm/min), and the displacement load curve of each sample was recorded. A micrometer was used to measure the lateral displacement of the sample and we calculated Poisson's ratio. The elastic modulus of the sample was measured by an elastic modulus meter (Grindosonic, Belgium). The shear frequency and torsional frequency of the sample under impact were measured by the torsional modulus instrument (Grindosonic, Belgium), and the torsional modulus of the sample was obtained by calculation. Figure 6 shows the test instruments.

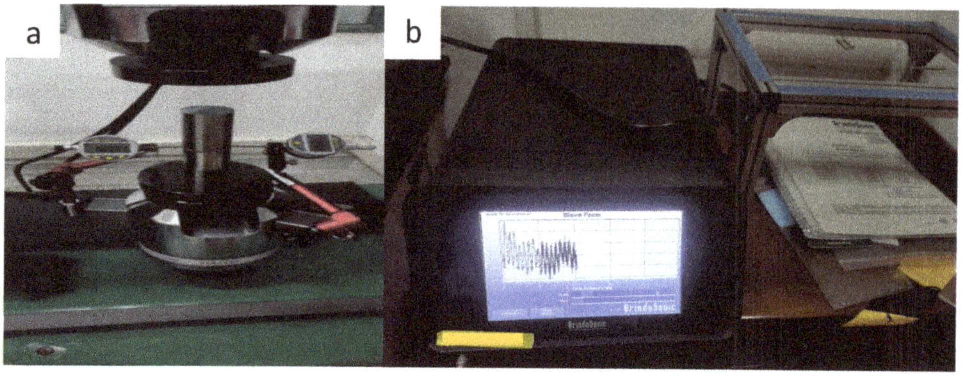

Figure 6. (**a**) Universal testing machine and micrometer, (**b**) elastic modulus instrument.

4. Results and Discussion

4.1. Density

The sintering pictures of composite discs with different fractal structures are shown in Figure 4b. It can be observed that, no matter what kind of fractal structure is used, the composite keeps the disc shape. Figure 7 shows the density of the sintered aluminum matrix composites with different fractal structures. For comparison, a pure aluminum disk was made by SPS in the same way, and the density of the aluminum was known (2.74 g/cm^3) [25]. We compared the densities of the different samples and found that the density of the aluminum alloy without a fractal structure was 2.7292 g/cm^3, which is close to the theoretical density, indicating that the sintering regime is reliable, while the densities of the samples with fractal structures were higher than that without a fractal structure, which is due to the fact that the density of alumina ceramic (3.62 g/cm^3) is greater than that of pure aluminum (2.73 g/cm^3). The differences between the densities of the different samples, are due to the mass of the fractal structure used. We calculated the theoretical density of each sample based on the mass fraction of alumina ceramics, and found that the actual densities were above 98% of the theoretical density values, indicating that the fabricated composites were densified and reliable in performance.

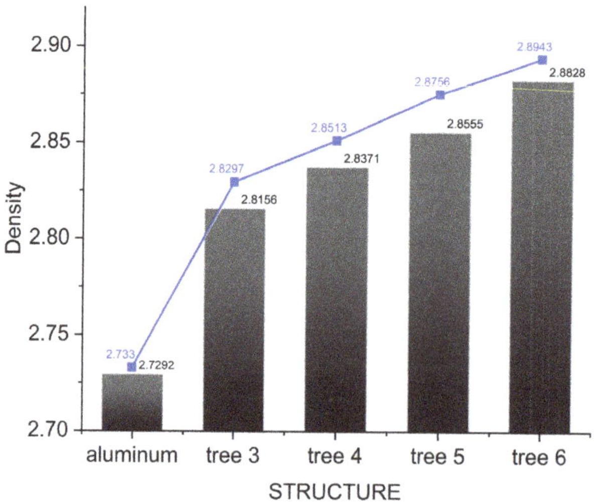

Figure 7. The density of the sintered aluminum alloy composites with different fractal structures. The blue line represents the theoretical density of the composite.

4.2. Microstructure

Figure 8 shows SEM images of the surface and cutting surface of the fabricated alumina ceramic/aluminum composite. Figure 8a shows an SEM image of the surface of the pure aluminum phase. The white substance is the lumpy precipitation phase of aluminum powder when sintering. The production of the precipitated phase is related to the particle deformation caused by the pressure and heating temperature. The pressure in the sintering process causes a small amount of Fe, Cu, Mg, and other elements in the aluminum powder, to precipitate out. When heated, these elements have a high melting point and do not easily form a liquid phase. The black part is a pore that is not closed by sintering shrinkage. Figure 8b–d shows SEM images of the cutting surface of the composite material, showing the interface between the Al_2O_3 ceramic and the aluminum matrix, observed at different magnifications. Figure 8b shows the interface phase at a resolution ratio of 400 μm. It can be observed that there is no porosity and impurity arrangement on the interface of the alumina ceramics, and the structure is tight and the porosity is low. Figure 8c,d shows the interface at higher resolutions. At these resolutions, we can observe that ceramic particles are distributed in the 3D printed ceramic lattice, with a maximum particle size of about 20μm. After sintering, the porous ceramic lattice has a rough and inhomogeneous lattice organization due to the presence of a few pores. At the same time, a large number of protrusions and depressions are generated, due to the cracking of the photosensitive resin filled between the particles at high temperatures, which is responsible for the uneven surface of the ceramic phase. It is known that at 800 K, since Mg is more reactive than Al, Mg will react with O in Al_2O_3 to form MgO, therefore, the Mg atoms in the aluminum alloy matrix will react with Al_2O_3 at the interface [26]:

$$3Mg + Al_2O_3 \rightarrow 3MgO + 2Al$$

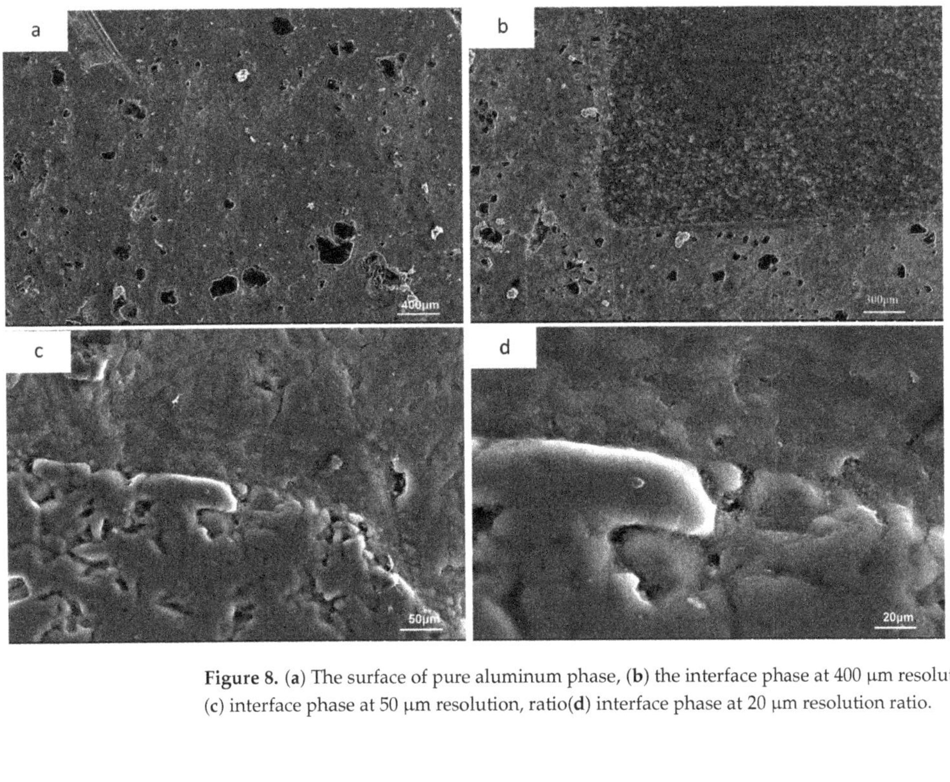

Figure 8. (**a**) The surface of pure aluminum phase, (**b**) the interface phase at 400 μm resolution ratio (**c**) interface phase at 50 μm resolution, ratio(**d**) interface phase at 20 μm resolution ratio.

Mg atoms in the aluminum matrix are continuously dispersed to the interface and ceramic phase, and the O atoms of Al_2O_3 are dispersed to the aluminum matrix. Therefore, the reaction products at the interface mainly contain Al_2O_3, MgO, and Al. The interfacial reaction, and the generation of new substances, will produce a transition layer between the alumina ceramics and the aluminum metal matrix, which makes the ceramic phase and metal closely bonded. Therefore, we observed that there is no gap at the interface between the ceramic phase and the aluminum matrix, and there is not a large number of pores and impurities.

Figure 9a shows an SEM image of the cutting surface at a resolution of 10 μm. Figure 9b–d shows the EDS distribution of the elements Al, O, and Mg, respectively. It can be seen from the element distribution diagram that Mg is enriched on the aluminum matrix side, while the content is less on the alumina ceramic side. The distribution of O is more toward the alumina ceramic side and less on the aluminum matrix side. This shows that Mg in the aluminum matrix disperses continuously to the interface and ceramic phase, while O atoms of Al_2O_3 disperse to the aluminum matrix, and the existence of an interface transition layer during sintering.

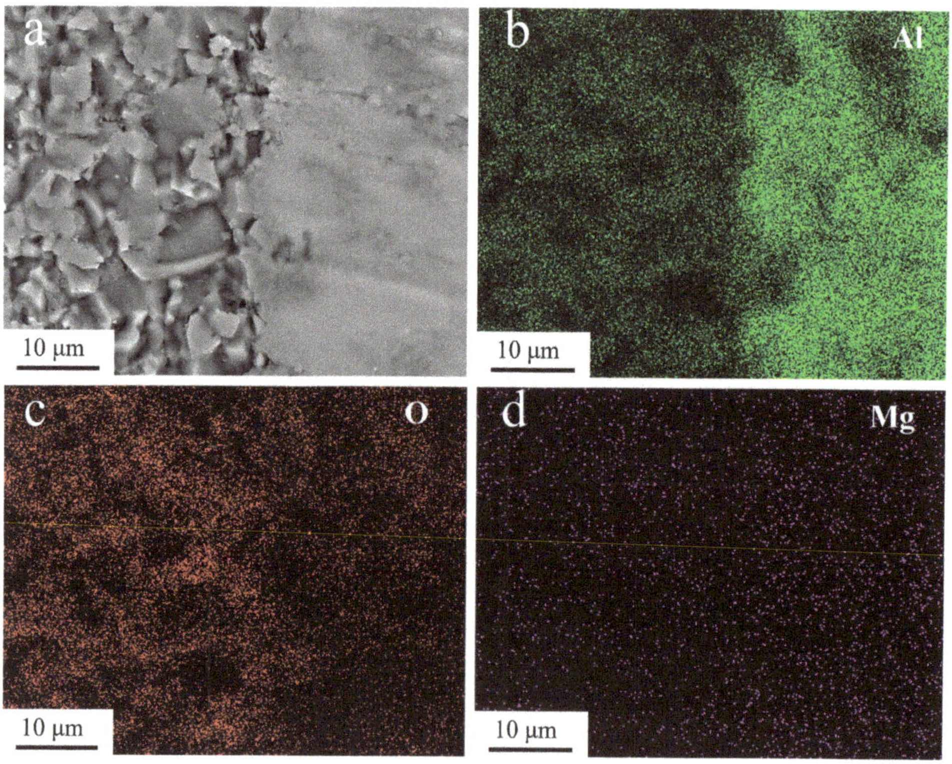

Figure 9. (**a**) SEM images of cutting surface at 10 μm resolution, (**b**) aluminum distribution map, (**c**) oxygen distribution diagram, (**d**) magnesium distribution map.

4.3. Compression Strength

In this experiment, we put the pure aluminum and aluminum matrix composite discs under the universal testing machine, and recorded the displacement–load curves of the samples. At the same time, the elastic modulus of each sample was measured, by an elastic modulus meter. Since too much load will make the deformation of the sample enter the stage where elastic deformation and plastic deformation act together, the maximum load applied to each sample was 90 KN.

Figure 10a shows the displacement load curves of the pure aluminum sample and samples with fractal structures, and the maximum load applied to each sample is 90 KN. As can be seen from the figure, the fluctuation of the displacement–load curve is relatively small and the curve is smooth, indicating that the whole sample has no structural damage during the compression test. For aluminum matrix composites with powders as structural units, the mechanical properties depend mainly on the bonding state between the powders. The results show that the bond between the powders in each sample is good. Table 3 shows the maximum displacement of each sample under the maximum load, and comparing the relative density of each sample and the mass fraction of Al_2O_3 ceramics.

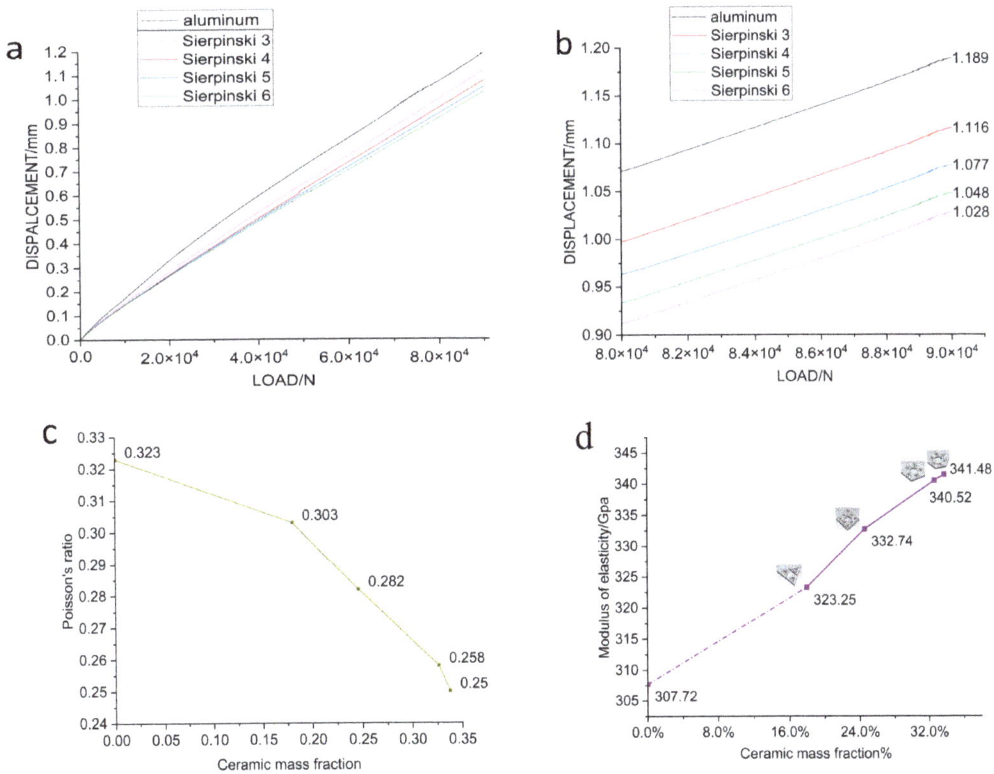

Figure 10. (**a**) Displacement–load curves of samples, (**b**) displacement–load curves of the disks with the four fractal structures and the pure aluminum disk, under high load (80–90 MPa), (**c**) relationship between the Poisson's ratio of the sample and the mass fraction of the ceramics in the sample, (**d**) relationship between the elastic modulus and the mass fraction of the ceramics in the sample.

Table 3. Relationship between maximum displacement and ceramic content.

	Sierpinski				**Al**
	3	**4**	**5**	**6**	
Density/(g/cm^3)	2.8504	2.9047	2.9675	2.9763	2.7292
Mass fraction	17.98%	24.57%	32.63%	33.75%	0
Maximum displacement/mm	1.1158	1.0772	1.0483	1.0277	1.18894

Figure 10b shows the displacement–load curves for the samples containing the four fractal structures, as well as the pure aluminum sample, under high loads (80–90 MPa). Figure 10c shows the relationship between the Poisson's ratio of the sample and the mass fraction of the ceramics in the sample. It can be seen from the table and figure that the addition of a fractal structure reduces the displacement of the composite under the maximum load to varying degrees, among which, the fractal structure of Sier 6 has the greatest influence, reducing the maximum displacement by 0.16124 mm, while the fractal structure of Sier 3 has the least influence, where the maximum displacement is only reduced by 0.07314 mm. With an increase in the ceramic content in the sample, the maximum displacement of the sample under high load gradually decreased with the use of different fractal structures, and the reduction ratio of displacement was between 6.15% and 13.56%. We hypothesize that the improvement in the compressive strength of the sample was positively correlated with the ceramic mass fraction of the alumina containing ceramics. To verify our conjecture, we tested the elastic modulus of each sample and plotted the relationship between the elastic modulus and the ceramic mass fraction (Figure 10d).

As can be seen from Figure 10d, with an increase in the mass fraction of alumina ceramics in the sample, the elastic modulus of the sample also increases, reaching a maximum of 341.48 GPa, with the increase of the elastic modulus ranging between 5.04 and 10.97%, indicating that the addition of an alumina ceramic fractal structure improves the compressive performance of the composite. The influence of each fractal structure on the compressive resistance is consistent with the mass proportion of the structure. The mass fraction of alumina ceramic in the Sier 3 structure is 17.98%, and its increase in the elastic modulus of the sample is 5.04%. The mass fraction of the alumina ceramic in the Sier 4 structure is 24.57%, and the increase in the elastic modulus of the sample is 8.13%. The mass fraction of alumina ceramic in Sier 5 structure is 32.63%, and its increase in the elastic modulus of the sample is 10.65%. The mass fraction of alumina ceramic in the Sier 6 structure is 33.75%, and its increase in the elastic modulus of the sample is 10.97%. According to previous studies, Zamani et al. [27] prepared nano-Al_2O_3 particle-reinforced aluminum matrix composites by a traditional powder metallurgy method, and analyzed the effects of the Al_2O_3 content on the microstructure and mechanical properties of the aluminum matrix composites. The results showed that the hardness and compressive strength of the composites were improved with the increase in Al_2O_3 content. Nassar et al. [28] studied the wear and mechanical properties of aluminum matrix composites reinforced by nano-TiO_2 particles with different contents, and obtained similar strengthening laws. The results of the above studies suggest that our result is reliable.

Through the previous part of research and literature reading, we concluded that the main reasons for the improvement in the compressive strength of composite materials containing an alumina ceramic fractal structure are as follows:

(1) The addition of a fractal structure of alumina ceramics reduces the defects in the aluminum matrix. As can be seen from Figure 8a, there are defects such as pores and precipitates on the surface of the aluminum matrix, while alumina ceramics have fewer internal defects, due to solid sintering. After the ceramic structure is added, the alumina ceramic is evenly distributed in the aluminum alloy matrix, which effectively reduces the porosity and defects in the sample (Figure 11), hinders the plastic deformation of the aluminum alloy matrix, and is conducive to the improvement of the compressive strength.

(2) Al_2O_3 in the ceramic matrix reacts with Al and Mg in the aluminum matrix to form a transition layer. The transition layer connects the aluminum alloy matrix to the ceramic structure, enhances the interface bonding, and promotes the load transition between the two. The robust Al_2O_3–Al interface can effectively carry out load transfer, thus delaying the occurrence of interface depolymerization. When the matrix is under pressure, Al_2O_3 will play the role of crack bridging, due to the strong bonding of the Al–Al_2O_3 interface. Not only that, but the transition layer acts as a dense spherical shell that protects the ceramic structure from damage. Therefore, the ceramic fractal

structure can maintain the structural integrity under large compressive loads and further hinder the occurrence of displacement.

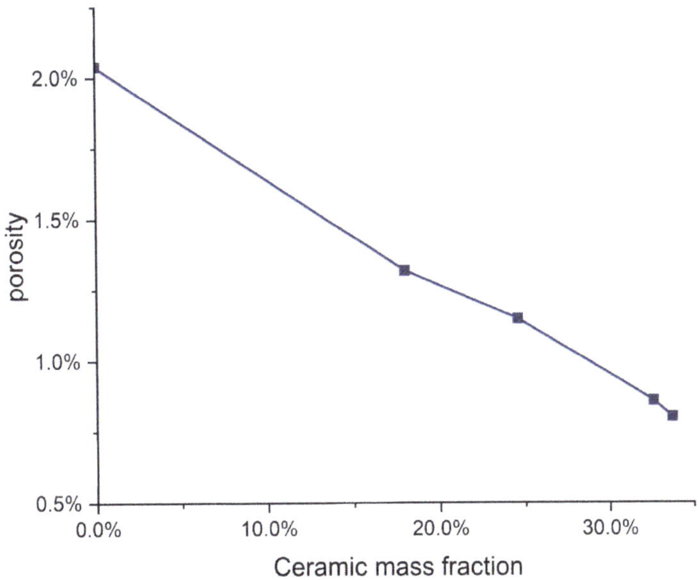

Figure 11. The relationship between porosity and mass fraction of ceramics.

The fractal ceramic structures used in this study have ordered structures, while the distribution of ceramic phases in traditional ceramic reinforced composites is homogeneous and disordered. In order to compare the effects of ordered and disordered structures on the mechanical properties of composites, we compared the elastic modulus of alumina ceramic/aluminum composites prepared by adding alumina ceramic powder.

Figure 12 shows the comparison of elastic modulus. The two curves represent the elastic modulus of the alumina ceramic composite reinforced with a fractal structure in this experiment, and the elastic modulus of the alumina ceramic/aluminum composite made with alumina ceramic powder. By comparing the elastic modulus of the composites made by the two methods, with the same ceramic content, we found that the elastic modulus of the composites strengthened with fractal structure ceramics are larger, indicating that their compression performance is better. The reason is, that the fractal ceramic structures used in this paper are ordered structures, while in the composite made of alumina ceramic powder, the ceramic phases are scattered and disordered. Wan et al. [29], for example, made alumina/aluminum composites with a pearl-like microlayer structure. They found that the distribution of oxygen elements in the composite was uniform, which proved that alumina was uniformly distributed in the aluminum matrix. Chen et al. [30] prepared a new type of porous aluminum alloy composite and also found that the distribution of alumina in the composite was uniform. This proves that ceramic phases in composites made of alumina ceramic powders are scattered and disordered.

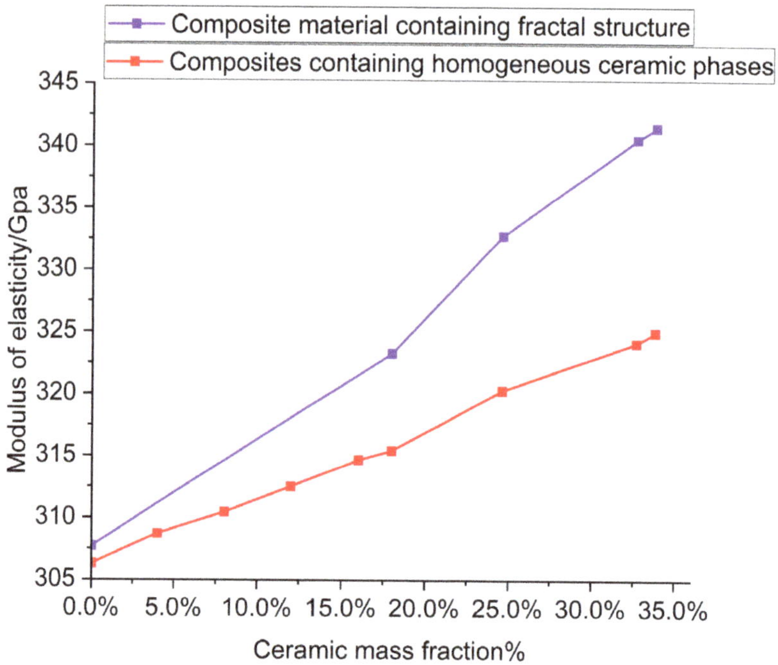

Figure 12. The relationship between elastic modulus and ceramic content of alumina ceramic/aluminum composites made by two different methods.

From the results of the elastic modulus, we can see that the addition of ordered structure has a greater improvement on the compressive properties of composites, than that of disordered structure. This is due to the structural characteristics of ordered structures, and the fact that ordered structures have fewer defects after solid sintering, before making composites.

4.4. Torsional Strength

As one of the important mechanical properties of metal and its composite materials, torsional strength has not attracted much attention in the previous research on composite materials. In order to test the influence of ceramic fractal structure on the torsional properties of aluminum, we used the torsional modulus instrument to measure the shear frequency and torsional frequency of the sample under impact force, and then obtained the torsional modulus, G, of the sample by calculation. G = 116.29 GPa for the pure aluminum sample. Table 4 shows the torsional modulus of each composite and the mass fraction of the fractal structure in each composite.

Table 4. Torsional modulus and ceramic content.

		Mass Fraction of Fractal Structure	*G* (GPa)
Sierpinski	3	17.98%	124.26
	4	24.57%	129.53
	5	32.63%	135.47
	6	33.75%	136.50
Al		0	116.03

According to the data obtained, the relationship between the mass fraction of the fractal structure and the torsional modulus of the sample is plotted (Figure 13). It can be seen from Figure 13 and Table 4 that the torsional modulus of the sample increases with the addition of the fractal structure, indicating that the torsional properties of the sample are enhanced, and the torsional modulus of the sample increases from 6.66% to 17.45%. Moreover, with the increase in the mass fraction of the fractal structure contained in the sample, the torsional modulus of the sample also gradually increases.

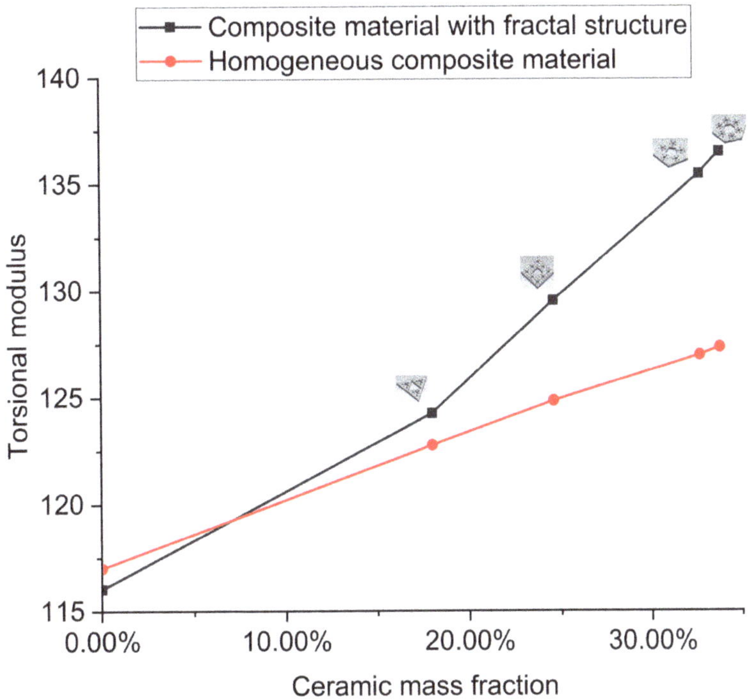

Figure 13. Comparison of torsional modulus.

We know that, one of the most important toughening mechanisms for ceramic/metal composites is the crack bridging of the ductile tube ligament, in which an unbroken metal layer straddles the crack wake and then pulls out [31]. The unfractured metal toughens the composite, by resisting crack opening displacement. When the metal ligament finally breaks, the ceramic begins to pull out of the metal phase, causing frictional sliding, which dissipates the strain energy [32]. In addition, the "multiple cracking" fracture mode also contributes to the cracking resistance of the composite [33]. Multiple cracks reduce crack tip stress, due to greater damage distribution and higher energy absorption, resulting in higher toughness. Le et al. [34] studied the high-cycle fatigue behavior of three kinds of cast aluminum alloys with different microstructure characteristics, under axial, torsional, and proportional tension-torsional loading conditions. Under different loading conditions, the fracture of all failure samples was caused by defects. Under axial loading, the location of initiation defects on the whole section is random, while under torsional loading, the location of initiation defects is near the surface. This is due to the difference in volume at high stresses, under different loading conditions. The entire section is in a high stress state when loaded axially, while there is a stress gradient when loaded torsionally, and only the outer region is in a high stress state [35].

The principle of adding fractal structure ceramics to improve the torsional performance is as follows: with the addition of the parting structure, the porosity of the original aluminum matrix is reduced to a certain extent, the number and distribution of internal defects are improved, and the torsional performance is improved. In addition to the crack mode, the influence of the interface strength is also significant. It is well known that interface strength plays a key role in the toughening of ceramic/metal composites [36]. At the same time, due to the different mass fraction of the fractal structure, the contact area between it and the aluminum matrix is also different. The larger the mass fraction of the fractal structure, the more complete the contact between it and the aluminum matrix. As described in the previous chapter, the interface between the ceramic phase and the aluminum metal creates a layer of transition, which enhances the interface bonding and promotes the load transition between the two. Therefore, the larger the mass fraction of the fractal structure, the stronger the interface bonding. It is known that crack initiation occurs on the surface farthest from the center of rotation, under torsional fatigue [37]. The annular features are formed due to shear stress and abrasion between the two separated surfaces. These annular features converge towards the central region of rotation, where the final rupture occurs. In the final fracture zone, equiaxial dimples appear, and they are formed due to local ductile fracture of the material [38]. Therefore, the existence of the transition layer between the ceramic phase and the metal phase can hinder the crack development and promote the deflection or bridging of the crack [39]. The crack expands in the Al grain and terminates in the adjacent Al_2O_3 grain, showing obvious crack passivation characteristics [40]. At the same time, the transition layer may cause crack propagation and deflection, thus consuming more energy [41]. Since crack initiation occurs on the surface farthest from the rotation center, the transition layer of a fractal structure with a larger mass fraction is in contact with the crack earlier than that of a fractal structure with a smaller mass fraction, that is, it has a greater impact on crack growth and load transfer. This explains why the addition of a fractal structure with a larger mass fraction, can lead to a greater improvement in the torsional properties of the sample.

We also calculate the torsional modulus of the composite with a homogeneous ceramic phase at the same ceramic content. Both groups of curves tend to be linear, indicating that the addition of the ceramic phase uniformly enhances the torsional properties of the composites. Under the same ceramic content, the torsional modulus of the composite with a fractal ceramic structure is greater than that of the composite with a homogeneous ceramic phase. The results show that the fractal structure is more effective than the homogeneous ceramic phase in enhancing the torsional properties of composites, and the ordered structure is more effective than the disordered structure in enhancing the torsional properties of composites.

By comparing the effects of fractal structures on the elastic modulus and torsional modulus, it is found that the influence of fractal structures on the torsional properties of composites is greater than that on their compressive properties, when the composites contain the same fractal structures.

5. Conclusions

In this paper, four kinds of fractal structures: triangle, square, pentagon, and hexagon, were designed and fabricated. Using SPS technology, the ceramic fractal structures and aluminum powder were fabricated into aluminum matrix composites, with certain mechanical properties. The following conclusions were obtained by testing various mechanical properties of the ceramic fractal structures.

(1) The results of SEM and elemental analyses show that the addition of a fractal structure reduces the defects of the aluminum matrix, and the interface reaction produced by sintering will produce a transition layer between the alumina ceramic and aluminum matrix, so that the ceramic phase and metal bond closely.
(2) The addition of a ceramic fractal structure can improve the compressive and torsional properties of composite materials. The increase range of the elastic modulus is 5.04–10.97%; the increase in the torsion modulus is 10.65–34.97%.

(3) A fractal ordered structure enhances the mechanical properties of composites more than a homogeneous structure. When the composite materials contain the same fractal structure, the influence of the fractal structure on the torsional properties is greater than that on the compression properties.

Author Contributions: Conceptualization, X.Z. and J.Z.; experimental design, X.Z. and J.Z.; testing of equipment, Q.J. and J.S.; writing—original draft preparation, X.Z.; writing—review and editing, Q.J. and J.S. All authors have read and agreed to the published version of the manuscript.

Funding: The authors greatly acknowledge the financial support from the National Natural Science Foundation of China (grant no. 43200021).

Institutional Review Board Statement: Not applicable.

Informed Consent Statement: Not applicable.

Data Availability Statement: Not applicable.

Conflicts of Interest: The authors declare that they have no known competing financial interest or personal relationship that could have appeared to influence the work reported in this paper.

References

1. Doorbar, P.J.; Kyle-Henney, S. Development of Continuously-reinforced Metal Matrix Composites for Aerospace Applications. In *Comprehensive Composite Materials II*; Beaumont, P.W.R., Zweben, C.H., Eds.; Elsevier: Amsterdam, The Netherlands, 2018; Volume 2, pp. 439–463.
2. Liao, Z.; Abdelhafeez, A.; Li, H.; Yang, Y.; Diaz, O.G.; Axinte, D. State-of-the-art of Surface Integrity in Machining of Metal Matrix Composites. *Int. J. Mach. Tool Manufact.* **2019**, *143*, 63–91. [CrossRef]
3. Hayat, M.D.; Singh, H.; He, Z.; Cao, P. Titanium Metal Matrix Composites: An Overview. *Compos. Appl. Sci. Manuf.* **2019**, *121*, 418–438. [CrossRef]
4. Hao, S.M.; Mao, J.W.; Xie, J.P. Research and Development of Ceramic Particle Reinforced Aluminum Matrix Composites. *Powder Metall. Ind.* **2018**, *281*, 56–62.
5. Yang, L.; Pu, B.; Zhang, X.; Sha, J.; He, C.; Zhao, N. Manipulating Mechanical Properties of Graphene/Al Composites by an in-situ Synthesized Hybrid Reinforcement Strategy. *J. Mater. Sci. Technol.* **2022**, *123*, 13–25. [CrossRef]
6. Xi, L.; Guo, S.; Ding, K.; Prashanth, K.; Sarac, B.; Eckert, J. Effect of Nanoparticles on Morphology and Size of Primary Silicon and Property of Selective Laser Melted Al-high Si Content Alloys. *Vacuum* **2021**, *191*, 110405. [CrossRef]
7. Jiang, B.; Tan, W.; Bu, X.; Zhang, L.; Zhou, C.; Chou, C.C.; Zhonghao, B. Numerical, Theoretical, and Experimental Studies on the Energy Absorption of the Thin-walled Structures with Bio-inspired Constituent Element. *Int. J. Mech. Sci.* **2019**, *164*, 105173. [CrossRef]
8. Zhang, L.; Bai, Z.; Bai, F. Crashworthiness Design for Bio-inspired Multi-cell Tubes with Quadrilateral, Hexagonal and Octagonal Sections. *Thin-Walled Struct.* **2018**, *122*, 42–51. [CrossRef]
9. Wang, Z.; Zhang, J.; Li, Z. On the Crashworthiness of Bio-inspired Hexagonal Prismatic Tubesunder Axial Compression. *Int. J. Mech. Sci.* **2020**, *186*, 105893. [CrossRef]
10. Mandelbrot, B. How long is the coast of Britain Statistical Self-similarity and Fractional Dimension. *Science* **1967**, *156*, 636–638. [CrossRef]
11. Farina, I.; Goodall, R.; Hernández-Nava, E.; di Filippo, A.; Colangelo, F.; Fraternali, F. Design, Microstructure and Mechanical Characterization of Ti_6Al_4Vreinforcing Elements for Cement Composites with Fractal Architecture. *Mater. Des.* **2019**, *172*, 107758. [CrossRef]
12. Wang, G.; Gu, Y.; Zhao, L.; Xuan, J.; Zeng, G.; Tang, Z.; Sun, Y. Experimental and Numerical Investigation of Fractal-tree-like Heat Exchanger Manufactured by 3D Printing. *Chem. Eng. Sci.* **2019**, *195*, 250–261. [CrossRef]
13. Li, Z.; Ma, W.; Xu, P.; Yao, S. Crashworthiness of Multi-cell Circumferentially Corrugated Square Tubes with Cosine and Triangular Configurations. *Int. J. Mech. Sci.* **2020**, *165*, 105205. [CrossRef]
14. Rian, I.; Sassone, M. Tree-inspired Dendriforms and Fractal-like branching structures in Architecture: A brief historical overview. *Front. Archit. Res.* **2014**, *3*, 298–323. [CrossRef]
15. McKittrick, J.; Chen, P.; Tombolato, L.; Novitskaya, E.; Trim, M.; Hirata, G.A.; Olevsky, E.; Horstemeyer, M.; Meyers, M. Energy Absorbent Natural Materials and Bioinspired Design Strategies: A review. *Mater. Sci. Eng. C.* **2010**, *30*, 331–342. [CrossRef]
16. Revilla-Leon, M.; Ozcan, M. Additive Manufacturing Technologies Used for Processing Polymer, Current Status and Potential Application in Prosthetic Dentistry. *J. Prosthodont.* **2019**, *28*, 146–158. [CrossRef]
17. Viccica, M.; Galati, M.; Calignano, F.; Iuliano, L. Design, Additive Manufacturing, and Characterisation of a Three-dimensional Cross-based Fractal Structure for Shock Absorption. *Thin-Walled Struct.* **2022**, *181*, 110106. [CrossRef]
18. Wu, J.; Zhang, Y.; Zhang, F.; Hou, Y.; Yan, X. A Bionic tree-like Fractal Structure as Energy Absorber under Axial Loading. *Eng. Struct.* **2021**, *245*, 112914. [CrossRef]

19. Prasad, S.; Verma, S. Fractal Interpolation Function on Products of the Sierpiński Gaskets. *Chaos Solit. Fract.* **2023**, *166*, 166988. [CrossRef]
20. Liang, Y.; Xiang, S.; Li, T.; Zhang, X. Ultrafast Fabrication of High-density Al–12Si Compacts with Gradient Structure by Electro-discharge Sintering. *J. Manuf. Process.* **2020**, *54*, 301–308. [CrossRef]
21. Chak, V.; Chattopadhyay, H.; Dora, T.L. A Review on Fabrication Methods, Reinforcements and Mechanical Properties of Aluminum Matrix Composites. *J. Manuf. Process.* **2020**, *56*, 1059–1074. [CrossRef]
22. Wang, D.; Wang, K.; Wang, X. Ultrafine-grained Twinning-induced Plasticity Steel Prepared by Mechanical alloying and Spark Plasma sintering. *Mater. Sci. Eng. A.* **2022**, *830*, 142302. [CrossRef]
23. Sweet, G.A.; Brochu, M.; Hexemer, R.L.; Donaldson, I.W.; Bishop, D.P. Microstructure and Mechanical Properties of Air Atomized Aluminum Powder Consolidated Via Spark Plasma Sintering. *Mater. Sci. Eng. A.* **2014**, *608*, 273–282. [CrossRef]
24. Sadeghi, B.; Shamanian, M.; Ashrafizadeh, F.; Cavaliere, P.; Rizzo, A. Friction Stir Processing of Spark Plasma Sintered Aluminum Matrix Composites with Bimodal Micro- and Nano-sized Reinforcing Al_2O_3 particles. *J. Manuf. Process.* **2018**, *32*, 412–424. [CrossRef]
25. Venkataraman, S.; Haftka, R.; Sankar, B.; Zhu, H.; Blosser, M. Optimal Functionally Graded Metallic Foam Thermal Insulation. *AIAA J.* **2004**, *42*, 2355–2363. [CrossRef]
26. Levi, C.; Abbaschian, G.; Mehrabian, R. Interface Interactions During Fabrication of Aluminum Alloy-alumina Fiber Composites. *Metall. Trans. A* **1978**, *9*, 697–711.
27. Zamani, N.A.B.N.; Iqbal, A.K.M.A.; Nuruzzaman, D.M. Fabrication and Characterization of Al_2O_3 Nanoparticle Reinforced Aluminium Matrix Composite Via Powder Metallurgy. *Mater. Today Proc.* **2020**, *29*, 190–195. [CrossRef]
28. Nassar, A.E.; Nassar, E.E. Properties of Aluminum Matrix Nano Composites Prepared by Powder metallurgy Processing. *J. King Saud Univ.-Eng. Sci.* **2017**, *29*, 295–299. [CrossRef]
29. Hongbo, W.; Nathanael, L.; Urangua, J.; Eric, H.; Chaolin, W.; Qiang, L.; Hua-Xin, P.; Bo, S.; Tan, S. Fabrication and characterisation of alumina/aluminium composite materials with a nacre-like micro-layered architecture. *Mater. Design* **2022**, *223*, 111190.
30. Chena, J.; Liua, P.; Wangb, Y.; Songa, S.; Hou, H. Mechanical properties of a new kind of porous aluminum alloy composite from ceramic hollow spheres with high strength. *J. Alloys Compd.* **2022**, *910*, 164911. [CrossRef]
31. Marini, M.; Ismail, A. Torsional Deformation and Fatigue Behaviour of 6061 Aluminium Alloy. *IIUM Eng. J.* **2011**, *12*, 21–32. [CrossRef]
32. Lia, Y.; Retrainta, D.; Xueb, H.; Gaob, T.; Sun, Z. Fatigue Properties and Cracking Mechanisms of a 7075 Aluminum Alloy under Axial and Torsional Loadings. *Procedia Struct. Integr.* **2019**, *19*, 637–644. [CrossRef]
33. Koutiri, I.; Bellett, D.; Morel, F.; Augustins, L.; Adrien, J. High Cycle Fatigue Damage Mechanisms in Cast Aluminium Subject to Complex Loads. *Int. J. Fatigue* **2013**, *47*, 44–57. [CrossRef]
34. Le, V.; Morel, F.; Bellett, D.; Saintier, N.; Osmond, P. Simulation of the Kitagawa-Takahashi Diagram using a Probabilistic Approach for Cast Al-Si Alloys Under Different Multiaxial Loads. *Int. J. Fatigue* **2016**, *93*, 109–121. [CrossRef]
35. Le, V.; Morel, F.; Bellett, D.; Saintier, N.; Osmond, P. Multiaxial High Cycle Fatigue Damage Mechanisms Associated with the Different Microstructural Heterogeneities of Cast Aluminium Alloys. *Mater. Sci. Eng. A.* **2016**, *649*, 426–440. [CrossRef]
36. Trsko, L.; Guagliano, M.; Bokuvka, O.; Novy, F. Fatigue Life of AW 7075 Aluminium Alloy after Severe Shot Peening Treatment with Different Intensities. *Procedia Eng.* **2014**, *74*, 246–252. [CrossRef]
37. Zhang, J.; Shi, X.; Bao, R.; Fei, B. Tension–torsion High-cycle Fatigue Failure Analysis of $2A1_2$-T_4 Aluminum Alloy with Different Stress Ratios. *Int. J. Fatigue* **2011**, *33*, 1066–1074. [CrossRef]
38. Zhao, T.; Jiang, Y. Fatigue of 7075-T651 Aluminum Alloy. *Int. J. Fatigue* **2018**, *30*, 834–849. [CrossRef]
39. Zhang, J.; Shi, X.; Fei, B. High Cycle Fatigue and Fracture Mode Analysis of $2A1_2$-T_4 Aluminum Alloy under Out-of-phase Axial-Torsion Constant Amplitude Loading. *Int. J. Fatigue* **2012**, *38*, 144–154. [CrossRef]
40. Peng, J.; Jin, X.; Xu, Z.; Zhang, J.; Cai, Z.; Luo, Z.; Zhu, M. Study on the Damage Evolution of Torsional Fretting Fatigue in a 7075 Aluminum alloy. *Wear* **2018**, *402*, 160–168. [CrossRef]
41. Nourian-Avval, A.; Fatemi, A. Fatigue Life Prediction of Cast Aluminum Based on Porosity Characteristics. *Theor. Appl. Fract. Mech.* **2020**, *109*, 102774. [CrossRef]

materials

Article

Modification of Iron-Rich Phase in Al-7Si-3Fe Alloy by Mechanical Vibration during Solidification

Cuicui Sun [1], Suqing Zhang [1,*], Jixue Zhou [1,*], Jianhua Wu [1], Xinfang Zhang [1,2] and Xitao Wang [1,2]

1 Shandong Provincial Key Laboratory of High Strength Lightweight Metallic Materials, Advanced Materials Institute, Qilu University of Technology (Shandong Academy of Sciences), Jinan 250014, China
2 School of Metallurgical and Ecological Engineering, University of Science and Technology Beijing, Beijing 100083, China
* Correspondence: zhangsuqing1985@163.com (S.Z.); zhoujx@sdas.org (J.Z.); Tel.: +86-82608309 (S.Z.); +86-82608309 (J.Z.)

Abstract: The plate-like iron-rich intermetallic phases in recycled aluminum alloys significantly deteriorate the mechanical properties. In this paper, the effects of mechanical vibration on the microstructure and properties of the Al-7Si-3Fe alloy were systematically investigated. Simultaneously, the modification mechanism of the iron-rich phase was also discussed. The results indicated that the mechanical vibration was effective in refining the α-Al phase and modifying the iron-rich phase during solidification. The forcing convection and a high heat transfer inside the melt to the mold interface caused by mechanical vibration inhibited the quasi-peritectic reaction: L + α-Al_8Fe_2Si → (Al) + β-Al_5FeSi and eutectic reaction: L → (Al) + β-Al_5FeSi + Si. Thus, the plate-like β-Al_5FeSi phases in traditional gravity-casting were replaced by the polygonal bulk-like α-Al_8Fe_2Si. As a result, the ultimate tensile strength and elongation were increased to 220 MPa and 2.6%, respectively.

Keywords: modification mechanism; iron-rich phases; mechanical vibration; Al-7Si-3Fe alloy

check for updates

Citation: Sun, C.; Zhang, S.; Zhou, J.; Wu, J.; Zhang, X.; Wang, X. Modification of Iron-Rich Phase in Al-7Si-3Fe Alloy by Mechanical Vibration during Solidification. *Materials* **2023**, *16*, 1963. https://doi.org/10.3390/ma16051963

Academic Editor: Tony Spassov

Received: 24 December 2022
Revised: 10 February 2023
Accepted: 25 February 2023
Published: 27 February 2023

1. Introduction

Aluminum alloys are widely used in the transportation, aerospace, building, and packaging industries due to their lightweight properties, high specific strength, high corrosion resistance, and excellent recyclability [1–3]. In the aluminum industry, applying recycled raw materials reduces energy consumption by more than 95%. Recently, recycled aluminum alloys have been regarded as an alternative to crude aluminum under the background of carbon neutrality. Unfortunately, the Fe element is an inevitable and harmful impurity present in recycled aluminum alloys, and it is easily enriched during aluminum alloy recycling [4–6]. Various brittle iron-rich intermetallic phases such as Al_3Fe (θ), Al_8Fe_2Si (α), and Al_5FeSi (β) formed in the recycling process; among the plate-like or needle-shaped β-Al_5FeSi phase (β phase) not only split matrix but also promote the formation of casting defects, such as pore and shrinkage porosity. These can act as locations of high-stress concentrations and severely deteriorate the performance of the recycled aluminum alloys [7–9]. However, the iron-rich phase is also an ideal strengthening phase for recycled aluminum, which results from its high hardness, good wettability with the matrix, and thermal stability [10]. Therefore, modifying the iron-rich phases will extend the application of recycled aluminum alloys.

With regard to reducing the disadvantages of plate-like iron-rich phases, many researchers have been interested in modifications. It has been proved that the morphology of the iron-rich phase can be modified by melt treatment [5,11], neutralizing elements addition (such as Mn, La) [12,13], rapid solidification process [14,15], and melt superheating [16,17]. Thus, the plate-like iron-rich phases are replaced by Chinese-script, star-like, or polygonal bulk-like phases, which lead to an increase in UTS and elongation of recycled aluminum. However, the addition of neutralizing elements may form coarse phases, which could reduce the properties of the alloy. For example, Song et al. [12] point out that when the Mn content was higher than 0.93% in the Al-7Si-1.2Fe alloy, the plate-like Fe-rich phase was eliminated, but the size of the star-like Fe-rich phase increased, which reduced the ductility. The research [13] shows La additions can refine effectively the β-Fe phase in Al-7Si-4Cu-0.35Mg-0.2Fe, while exceeds 0.15 wt.%, the β-Fe and La-rich particles are crack initiation sources and mainly affect the final failure. In addition, neutralizing elements addition can also affect the recycling of aluminum alloys. Meanwhile, the applications of rapid solidification and melt superheating are also limited because of their higher costs and complicated processes.

It is well known that melt treatment is a relatively simple method in comparison with other processing and can effectively improve the properties of alloy [5,11,18]. During solidification, the melt is usually treated by mechanical vibration, ultrasonic waves, electromagnetic fields, and so on. Among them, mechanical vibration treatment has attracted more and more attention, thanks to its simple and low cost without introducing a new element. It has been proven that [19–23] mechanical vibration can refine grains due to the promotion of nucleation, reduce shrinkage porosities due to improving metal feeding, modify the second phase, and produce a more homogenous metal structure. These improved features can effectively enhance mechanical properties and lower susceptibility cracks. For example, after applying mechanical vibration to A356 alloy, the coarser dendrites transformed into fine and uniform equiaxed grains, and eutectic Si particles were refined, which increased the tensile strength, yield strength, elongation, and hardness by 35%, 42%, 63%, and 29%, respectively [20]. When the mechanical vibration range of 30~50 Hz is applied to LM25 aluminum alloy, the inner casting defects reduce and become smaller, and the scattering of mechanical properties is enhanced [21]. The flaky structure of aluminum transforms to fibrous by applying mechanical vibration of 100 HZ and 149 μm to Al-12Si alloy [22]. However, it is not clear whether mechanical vibration during solidification will modify the iron-rich phases in the Al-Si-Fe alloy.

The aim of the present work was to investigate the effect of mechanical vibration on the morphology and size of α-Al and the iron-rich phases of Al-7Si-3Fe. In addition, the mechanism of mechanical vibration on the morphology of the iron-rich phase was discussed by OM, SEM, XRD, the Thermo-Calc software, and so on.

2. Experimental

The raw materials were pure Al, Al-20 wt.%Si, and Al-20 wt.%Fe master alloy. The amount of pure Al, Al-20 wt.%Si, and Al-20 wt.%Fe master alloy were melted in a graphite crucible in an electric resistance furnace under 750 °C, and then the melt was degassed for 10 min with argon gas at 720 °C. After holding for 10 min at a temperature of 680 °C, the melt was poured into a stainless-steel mould, which was installed on a mechanical vibration device. The schematic diagram of the mechanical vibration device is shown in Figure 1. Subsequently, the mechanical vibration device was launched until the melt was totally solidified. The vibration conditions were frequency (30 Hz), amplitude (0.6 mm), and direction (vertical). The as-vibration cast Al-7Si-3Fe alloy was obtained (marked as as-vibration). For comparison, some of the melt was solidified under the same conditions without mechanical vibration, and the traditional cast Al-7Si-3Fe alloy was obtained (marked as as-cast). The chemical compositions of as-vibration and as-cast alloys were analyzed by an X-ray fluorescence (XRF) analyzer, as shown in Table 1.

Table 1. Chemical compositions of Al-Si-Fe alloys.

Alloy	Mass Fraction/%		
	Si	Fe	Al
As-cast	6.62	2.86	Bal.
As-vibration	6.47	2.87	Bal.

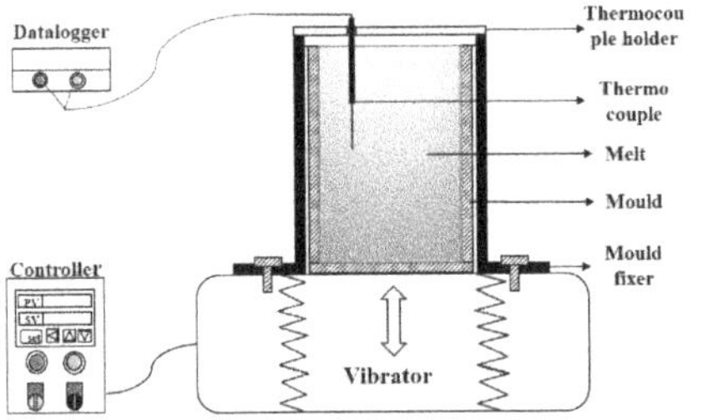

Figure 1. The schematic of a mechanical vibration device.

The solidification sequences of the Al-Si-Fe ternary system alloy were calculated using the TCAL4 database within the Thermo-Calc software with the assumption of equilibrium solidification under the condition of complete diffusion and redistribution of all solute atoms. The cooling curves of the alloy were measured by a thermocouple and datalogger. The thermocouple fixed by the holder was inserted into the melt during solidification and connected to the datalogger, which was used to record the data. Especially, the thermocouples were calibrated against the melting of pure aluminum, and the datalogger included a temperature measurement module and data recording software. Additionally, the cooling curves of the alloy were obtained based on the measured data.

Specimens for the metallographic examination were grinded through water matter paper, polished with 1.5 μm diamond grinding paste, and then etched with a solution of 1% hydrofluoric acid. Before observation of the three-dimensional morphology, a 10% sodium hydroxide solution was used to dissolve the Al matrix (0.5–1 h) in order to expose more of the Fe-phases embedded inside the Al matrix. The microstructures of Al-7Si-3Fe alloy were surveyed by a Zeiss Observer A1m optical microscope (OM), a Zeiss Evo MA10 scanning electron microscope (SEM, Carl Zeiss AG, Oberkochen, Germany) equipped with an Oxford X-Max 50 mm^2 energy dispersive spectroscope (EDS, Oxford Instruments, Oxford, UK), and a Smartlab 9Kw X-ray diffraction (XRD, Rigaku Corporation, Tokyo, Japan). The samples were machined into a tensile testing rod with a diameter of 5 mm and a gauge length of 25 mm. The tensile tests were carried out at room temperature using an MTS model E45 testing machine at a loading velocity of 1 mm/s. The ultimate tensile strength (UTS) and elongation to fracture are the average values of at least three individual repeated tests.

3. Results and Dicscussion

Figure 2a shows the XRD patterns of the Al-7Si-3Fe alloys. It is found that the as-cast and as-vibration Al-7Si-3Fe alloys consist of α-Al, Si, and iron-rich phases, especially the β-Al_5FeSi phases in the as-cast alloy, which are replaced by the α-Al_8Fe_2Si phases in the as-vibration alloy. In addition, it is obvious that the iron phases in the as-cast alloy are elongated, while those in the as-vibration alloy are polygonal, as shown in Figure 2b,c. Subsequently, the element contents of phases A and B in Figure 2b,c are analyzed by EDS, and the results are shown in Table 2. It is found that the Fe/Si ratio is about 1 in the plate-like phase (A) and 1.9 in the polygonal bulk-like phase (B). According to the literature [24], the Fe/Si ratio is about 1 for β-Al_5FeSi, and 1.9 for α-Al_8Fe_2Si in Al-Si-Fe alloys. Thus, it is confirmed that the plate-like phases in the as-cast alloy are β-Al_5FeSi and the polygonal bulk-like phase in the as-vibration alloy is α-Al_8Fe_2Si (comparing Figure 2b,c). According to references [8,25], the difference in morphology of the iron-rich phase is due to the β phase belonging to the monoclinic lattice crystal structure, while the α phase is a hexagonal lattice.

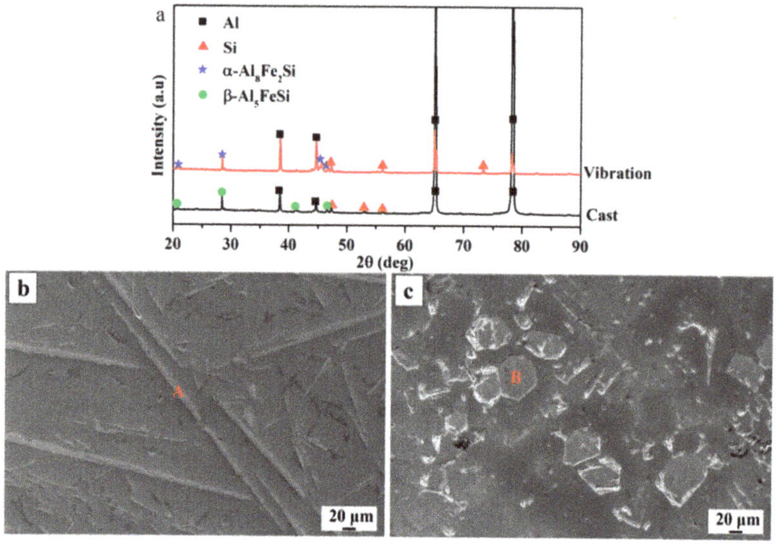

Figure 2. XRD (**a**) and SEM images of as-cast Al-7Si-3Fe alloy (**b**) and as-vibration (**c**).

Table 2. EDS results of the different iron-rich phases in Al-7Si-3Fe alloys.

Alloy	Phase	Atomic Percentage			Phase Formula
		Al	Si	Fe	
As-cast	A	71.02	15.69	13.29	β-Al_5FeSi
As-vibration	B	70.42	10.14	19.44	α-Al_8Fe_2Si

Figure 3 shows OM graphs of as-cast and as-vibration Al-7Si-3Fe alloys and the detailed three-dimensional shapes of the β-Al_5FeSi and α-Al_8Fe_2Si phases. As shown in Figure 3a, in the as-cast alloy, the microstructure consists of the developed α-Al dendrites and the thin and long plate-like β-Al_5FeSi phases. The later ones are distributed throughout the matrix, which split it and led to a severe decrease in properties. In the as-vibration alloy, the coarse α-Al dendrites are refined, and the coarse dendrites transform into a fine and uniform rose-like grain, while the thin and long plate-like β-Al_5FeSi phases are replaced by the polygonal, bulk-like α-Al_8Fe_2Si phases, as shown in Figure 3d. In addition, it is found that a small amount of short rod-like β-Al_5FeSi phase is in the matrix (Figure 3d). The polygonal bulk-like phase and short rod-like phase are mainly distributed in the grain junction or grain boundaries (Figure 3d), which reduces the disadvantage of the properties of the alloy. To further understand the iron-rich phase, the detailed three-dimensional shapes are shown in Figure 3b,c,e,f. It is visible that the β-Al_5FeSi phase exhibits a plate-like structure, with an average width of ~10 μm, especially, some plates cross each other and form a network structure, as Figure 3b,c. Dinnis et al. [26] has observed by serial sectioning that the needle-like β-Al_5FeSi phase in the 2D radiographs is actually plate-like and the plate-like phases form complex and interconnected network structure each other. After vibration, the 3D morphology of iron-rich is mainly a polyhedral structure, as shown in Figure 3e,f. According to Gao [27], the 3D morphology of the blocky α-AlFeSi particles indicates an obvious polyhedral shape from the corresponding fractography image. The tensile tests of as-cast and as-vibration alloys show that after applying vibration, the ultimate tensile strength and elongation increased from 190.6 MPa and 1.6% to 220 MPa and 2.6%, respectively.

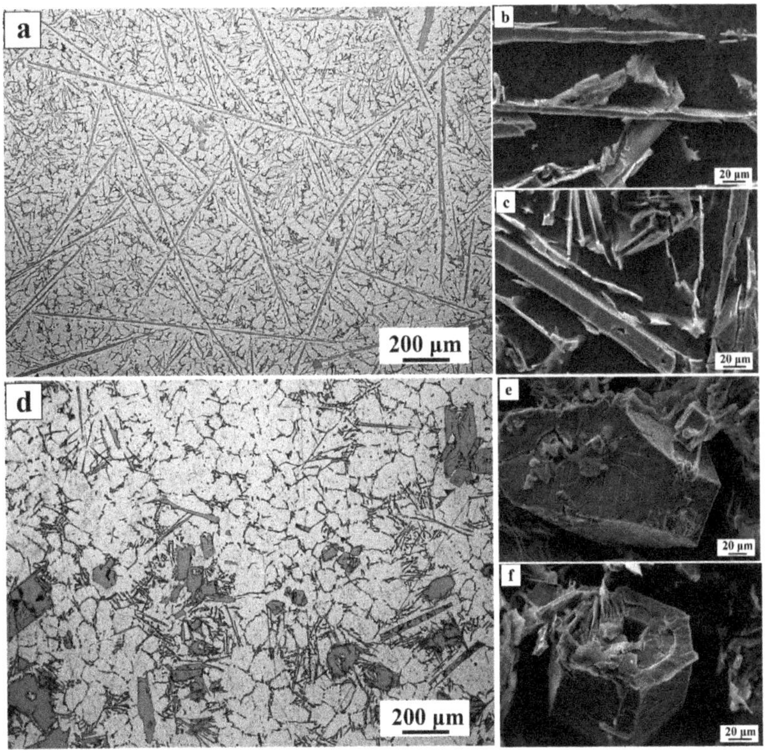

Figure 3. (**a**) OM of as-cast Al-7Si-3Fe alloy; (**b**,**c**) 3D morphologies of β-Al_5FeSi phase; (**d**) OM of as-vibration Al-7Si-3Fe alloy; (**e**,**f**) 3D morphologies of α-Al_8Fe_2Si phase.

Figure 4a indicates the solidification sequence of the Al-7Si-3Fe alloy based on Thermo-Calc software. The α-Al_8Fe_2Si phase first begins to crystallize at 658 °C, as shown in Figure 4a. As the temperature drops, the α-Al_8Fe_2Si phase continues to precipitate. When the temperature drops to 613 °C, the (Al) begins to form because of the decrease in Si and Fe in the liquid. As Si increases in the remaining melt and the temperature drops to 611 °C, the α-Al_8Fe_2Si phase begins to transform into the β-Al_5FeSi phase and a quasi-peritectic reaction occurs (L + α-Al_8Fe_2Si → (Al) + β-Al_5FeSi). When the temperature decreases to 574 °C, the liquid composition moves to the eutectic point, and the eutectic (Al), (Si), and β-Al_5FeSi phases crystallize until all of the liquid is exhausted. These solidification reactions are consistent with those reported in References [8,9]. Figure 4b shows the cooling curves of the as-cast and as-vibration alloys measured by experiment. It can be seen from Figure 4b that the as-vibration melt has a higher cooling rate, and the cooling rate played a critical role in the control of the solidification structure. Liu [14] indicated the coarse iron-rich intermetallics were refined to a significant extent by increasing the cooling rate. In addition, Figure 4b suggests the α-Al_8Fe_2Si phase has earlier precipitation time under vibration, compared with traditional casting.

First, the vibration promotes the formation of free crystals. According to crystal dissociating theory [28], some grains formed on the mold wall and the cooling liquid surface are more easily free to melt under the forced convection, which promotes the formation of grain. Additionally, it is well known that forced convection is induced by vibrations in the melt. Therefore, based on the solidification sequence in Figure 4a, the first precipitation α-Al_8Fe_2Si phases move toward melt in forced convection, and then the new grains continue to form in the mold wall and the cooling liquid surface. As a result, a large number of α-Al_8Fe_2Si phases formed in the melt, as shown in Figure 5a.

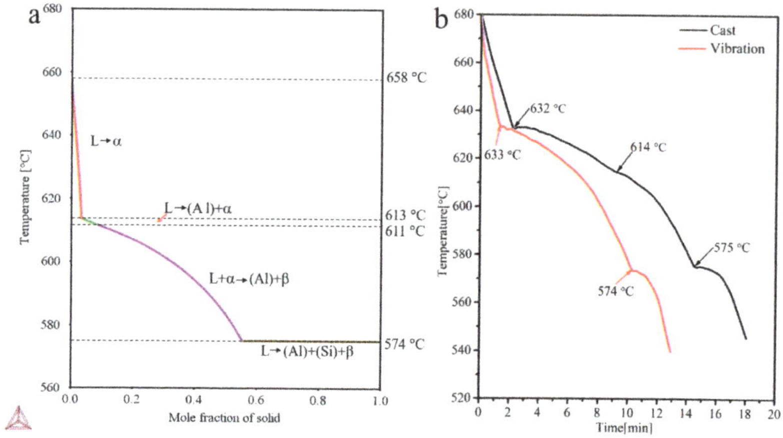

Figure 4. The solidification sequence of Al-7Si-3Fe based on Thermal-Calc software (**a**) and cooling curves of as-cast and as-vibration Al-7Si-3Fe alloys measured by experiment (**b**).

As mentioned above, the plate-like β-Al_5FeSi phases are substituted by the polygonal bulk-like α-Al_8Fe_2Si phases in an as-vibration alloy (as shown in Figures 2 and 3). The following three reasons are responsible for the modification mechanisms and the schematic drawing in Figure 5.

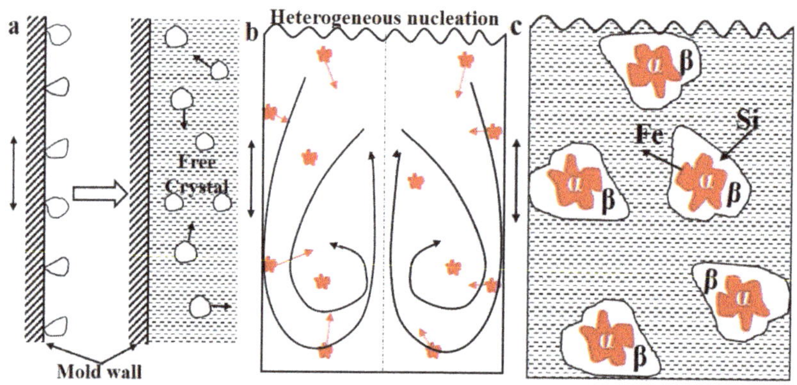

Figure 5. The schematic diagram of the transformation mechanism of iron-rich phase: (**a**) free crystal, (**b**) heterogeneous nucleation, and (**c**) element diffusion in quasi-peritectic reaction.

Second, the vibration promotes heterogeneous nucleation. References [28,29] show that forced convection caused by vibration can eliminate the difference in temperature and composition field in the melt, and the whole melt is in an undercooling state. In this case, a large number of effective nucleation sites can carry out heterogeneous nucleation in the melt, which increases the heterogeneous nucleation rate of the α-Al_8Fe_2Si phase (as shown in Figure 5b). In addition, Que et al. [30] reported that the possible heterogeneous nucleation substrates for the primary iron-rich intermetallic compounds are the casting mould wall and the native and in-situ oxides. Similarly, some research [31,32] also suggested that iron-rich intermetallic compounds can nucleate on the outer surface of oxide bi-films and xoides, by observation of direct contact. As we all know, the oxides are easily formed on the melt surface during the solidification process of aluminum alloy, and then the oxides on the melt surface are continuously introduced into the melt by vibration, thus promoting heterogeneous nucleation. Meanwhile, the oxides formed on the mould wall are washed down by vibration and move toward melting, which promotes nucleation.

Third, the vibration can improve the growth rate. As we all know, the vibration can induce a high heat transfer inside the molten metal to the mold interface, which leads to a high cooling rate of the melt, as shown in Figure 4b. A large number of α-Al_8Fe_2Si phase nuclei formed previously grow rapidly at a higher cooling rate. Reference [33] shows that when the α-Al_8Fe_2Si achieves to a certain size, it is difficult to transform to the β-Al_5FeSi through quasi-peritectic due to the hindrance of atomic diffusion (Figure 5c). However, not all α-Al_8Fe_2Si phase nuclei grow to a certain size rapidly under vibration; there are still some α-Al_8Fe_2Si phases that transfer to β-Al_5FeSi phases by quasi-peritectic reaction, such as some short rod-like β-Al_5FeSi phases in Figure 3d.

Above all, it is found that the quasi-peritectic reaction step of 614 °C disappears, as shown in Figure 4b, which indicates the quasi-peritectic reaction was hindered by vibration. Meanwhile, a lot of Fe and Si elements are consumed due to free crystals, and the high nucleation and growth rate of the α-Al_8Fe_2Si phase limit the eutectic reaction L → (Al) + β-Al_5FeSi + Si. At the same time, Figure 4b shows that the eutectic reaction time of 574 °C decreases under vibration. According to the solidification sequence (Figure 4a), the β-Al_5FeSi phase is formed mainly by quasi-peritectic reaction and eutectic reaction. As result, the plate-like β-Al_5FeSi phases are difficult to crystallize during solidification. However, there is still a small number of β-Al_5FeSi phases in the matrix, and these are refined to short rod-like, as shown in Figure 3d, due to the higher cooling rate under vibration. According to [34], the relationship between cooling rate and grain size is given by:

$$GS = a \times CR^{-b} \quad (1)$$

The value of b is in the range of 0.3–0.6. Thus, the calculated formula is:

$$GS = 34.7 \times CR^{-0.5} \quad (2)$$

Based on the above formula, the higher cooling rate can refine effectively grain. Meanwhile, the coarser α-Al dendrites are refined to a fine and uniform rose-like grain under vibration, as shown in Figure 3a,d. Therefore, applying vibration to the melt during solidification can effectively modify the iron-rich phase and refine the iron-rich phase and α-Al dendrites, which improve the properties of the alloy.

4. Conclusions

The modification behavior and mechanism of mechanical vibration on the iron-rich phase in an Al-7Si-3Fe alloy were investigated. The main findings are as follows:

(1) When the mechanical vibration (frequency: 30HZ, amplitude: 0.6 mm, direction: vertical) was applied on the Al-7Si-3Fe alloy during solidification, the coarse α-Al dendrites were refined to the fine and uniform rose-like grain, and the primary plate-like β-Al_5FeSi transformed to polygonal bulk-like α-Al_8Fe_2Si. The ultimate tensile strength and elongation were increased to 220 MPa and 2.6%.

(2) The solidification sequence of Al-7Si-3Fe alloy based on the Thermal-Calc was L → α-Al_8Fe_2Si (658 °C), L → α-Al_8Fe_2Si + (Al) (613 °C), quasi-peritectic reaction: L + α-Al_8Fe_2Si → β-Al_5FeSi (611 °C), and eutectic reaction: L → (Al) + β-Al_5FeSi + Si (574 °C)

(3) When applying vibration during solidification, the cooling rate of the Al-7Si-3Fe alloy increases, and the quasi-peritectic reaction step of 614 °C disappears. This is mainly due to a forced convection and a higher heat transfer caused by vibration.

Author Contributions: Conceptualization, S.Z. and J.Z.; methodology, C.S. and S.Z.; validation. J.Z.; formal analysis. C.S. and J.W.; investigation, C.S.; data curation, J.Z. and J.W.; writing – original draft preparation, C.S. and S.Z.; writing – review and editing, X.Z. and X.W.; supervision, X.W. All authors have read and agreed to the published version of the manuscript.

Funding: This research was funded by Youth Innovation and Technology Support Program of Shandong Provincial Colleges and Universities (2020KJA002) and the Shandong Province Key Research and Development Plan (2021SFGC1001).

Data Availability Statement: All data are available from the corresponding author on reasonable request.

Acknowledgments: The author (C. Sun) is grateful to Shandong Provincial Key Laboratory of High Strength Lightweight Metallic Materials at Shandong Advanced Materials Institute for providing an excellent working and study environment. All the authors wish to thank the financial support from the Youth Innovation and Technology Support Program of Shandong Provincial Colleges and Universities (Grant No. 2020KJA002), and the Shandong Province Key Research and Development Plan (Grant No. 2021SFGC1001) during the experiment.

Conflicts of Interest: The authors declare that they have no conflict of interest.

References

1. Jiao, X.; Liu, C.; Guoa, Z.; Tong, G.; Ma, S.; YYFZhang Xiong, S. The characterization of Iron-rich phases in a high-pressure die cast hypoeutectic aluminum-silicon alloy. *J. Mater. Sci. Technol.* **2020**, *51*, 54–62. [CrossRef]
2. Song, Z.; Magdysyuk, O.V.; Tang, L.; Sparks, T.; Cai, B. Growth dynamics of faceted $Al_{13}Fe_4$ intermetallic revealed by high-speed synchrotron X-ray quantification. *J. Alloys Compd.* **2021**, *861*, 158604. [CrossRef]
3. Mao, H.; Kong, Y.; Cai, D.; Yang, M.; Peng, Y.; Zeng, Y.; Zhang, G.; Shuai, X.; Huang, Q.; Li, K.; et al. β″ needle-shape precipitate formation in Al-Mg-Si alloy: Phase field simulation and experimental verification. *Comput. Mater. Sci.* **2020**, *184*, 109878. [CrossRef]
4. Song, D.; Zhao, Y.; Jia, Y.; Huang, G.; Zhang, Z.; Zhou, N.; Li, X.; Zheng, K.; Fu, K.; Zhang, W. Effect of B addition on the formation of Fe-rich phases in Al-Si-Fe alloys. *J. Alloys Compd.* **2023**, *930*, 167426. [CrossRef]
5. Cao, J.; Shuai, S.; Huang, C.; Hu, T.; Chen, C.; Wang, J.; Ren, Z. 4D synchrotron X-ray tomographic study of the influence of transverse magnetic field on iron intermetallic compounds precipitation behavior during solidification of Al-Si-Fe alloy. *Intermetallics* **2022**, *143*, 107471. [CrossRef]
6. Feng, S.; Liotti, E.; Lui, A.; Wilson, M.D.; Connolley, T.; Mathiesen, R.H.; Grant, P.S. In-situ X-ray radiography of primary Fe-rich intermetallic compound formation. *Acta Mater.* **2020**, *196*, 759–769. [CrossRef]
7. Chanyathunyaroj, K.; Patakham UKou, S. Limmaneevichitr, Microstructural evolution of iron-rich in scandium modified Al-7Si-0.3Mg alloys. *J. Alloys Compd.* **2017**, *692*, 865–875. [CrossRef]
8. Lee, S.; Kim, B.; Lee, S. Prediction of Solidification Paths in Al-Si-Fe Ternary System and Experimental Verification: Part I. Fe-Containing Hypoeutectic Al-Si Alloys. *Mater. Trans.* **2011**, *52*, 1053–1062. [CrossRef]
9. Chen, H.L.; Qing CH, E.N.; Yong, D.U.; Bratberg, J.; Engström, A. Update of Al-Fe-Si, Al-Mn-Si and Al-Fe-Mn-Si thermodynamic descriptions. *Trans. Nonferrous Met. Soc. China* **2014**, *24*, 2041–2053. [CrossRef]
10. Puncreobutr, C.; Phillion, A.B.; Rockett, P.; Horsfield, A.; Lee, P. In situ quantification of the nucleation and growth of Iron-rich intermetallics during Al alloy solidification. *Acta Mater.* **2014**, *79*, 292–303. [CrossRef]
11. Qiu, Y.; Li, X.; Huang, H.; Liu, M.; Xia, P.; Luo, Y.; Zhou, N. Effect of ultrasonic vobration on the microstructure and corrosion properties of the 7046 Al alloy. *Vacuum* **2022**, *205*, 111465. [CrossRef]
12. Song, D.; Zhao, Y.; Jia, Y.; Li, R.; Zhou, N.; Zheng, K.; Fu, Y.; Zhang, W. Study of the evolution mechanisms of Fe-rich phases in Al-Si-Fe alloys with Mn modification using synchrotron X-ray imaging. *J. Alloys Compd.* **2022**, *915*, 165378. [CrossRef]
13. Zhao, B.; Xing, S.; Shan, A.; Yan, G.; Jiang, X. Influence of La addition on Fe-rich intermetallic phases formation and mechanical properties of Al-7Si-4Cu-0.35Mg-0.2Fe alloys prepared by squeeze casting. *Intermetallics* **2023**, *153*, 107783. [CrossRef]
14. Liu, Y.; Luo, L.; Han, C.; Ou, L.; Wang, J.; Liu, C. Effect of Fe, Si and Cooling Rate on the Formation of Fe- and Mn-rich Intermetallics in Al–5Mg–0.8Mn Alloy. *J. Mater. Sci. Technol.* **2016**, *32*, 305–312. [CrossRef]
15. Becker, H.; Bergh, T.; Vullum, P.E.; Leineweber, A.; Li, Y. Effect of Mn and cooling rates on α-, β- and δ-Al–Fe–Si intermetallic phase formation in a secondary Al–Si alloy. *Materialia* **2019**, *5*, 100198. [CrossRef]
16. Haque, M.M.; Ismail, A.F. Effect of superheating temperatures on microstructure and properties of strontium modified aluminium–silicon eutectic alloy. *J. Mater. Process. Technol.* **2005**, *162–163*, 312–316. [CrossRef]
17. Wang, B.; Liu, X.; Wang, J.; Li, Q.; Liu, K.; Zhang, M. Uncovering the effects of Ce and superheat temeperature on Fe-rich intermetallic and microporosity formation in aluminum alloy. *Mater. Charact.* **2022**, *193*, 112226. [CrossRef]
18. Lin, C.; Wu, S.S.; Zhong, G.; Wan, L.; An, P. Effect of ultrasonic vibration on Fe-containing intermetallic compounds of hypereutectic Al–Si alloys with high Fe content. *Trans. Nonferrous Met. Soc. China* **2013**, *23*, 1245–1252. [CrossRef]
19. Taghavi, F.; Saghafian, H.; Haarrazi YH, K. Study on the effect of prolonged mechanical vibration on the grain refinement and density of A356 aluminum alloy. *Mater. Des.* **2009**, *30*, 1604–1611. [CrossRef]

20. Jiang, W.; Fan, Z.; Chen, X.; Wang, B.; Wu, H. Combined effects of mechanical vibration and wall thickness on microstructure and mechanical properties of A356 aluminum alloy produced by expendable pattern shell casting. *Mater. Sci. Eng. A* **2017**, *619*, 228–237. [CrossRef]
21. Varun, S.; Chavan, T.K. Influence of mould vibration on microstructural behaviour and mechanical properties of LM25 aluminium alloy using gravity die casting process. *Mater. Today Proc.* **2020**, *9*, 671. [CrossRef]
22. Abu-Dheir, N.; Khraisheh, M.; Saito, K.; Male, A. Silicon morphology modification in the eutectic Al–Si alloy using mechanical mold vibration. *Mater. Sci. Eng. A* **2005**, *393*, 109–117. [CrossRef]
23. Chirita, G.; Stefanescu, I.; Soares, D.; Silva, F.S. Influence of vibration on the solidification behaviour and tensile properties of an Al–18wt%Si alloy. *Mater. Des.* **2009**, *30*, 1575–1580. [CrossRef]
24. Mulazimoglu, M.H.; Zaluska, A.; Gruzleski, G.E.; Paray, F. Electron microscope study of Al-Fe-Si intermetallics in 6201 aluminum alloy. *Metall. Mater. Trans. A* **1996**, *27A*, 929–936. [CrossRef]
25. Roger, J.; Bosselet, F.; Viala, J.C. X-rays structural analysis and thermal stability studies of the ternary compound α-AlFeSi. *J. Solid State Chem.* **2011**, *184*, 1120–1128. [CrossRef]
26. Dinnis, C.M.; Taylor, J.A.; Dahle, A.K. As-cast morphology of iron-intermetallics in Al–Si foundry alloys. *Scr. Mater.* **2005**, *53*, 955–958. [CrossRef]
27. Gao, T.; Wu, Y.; Li, C.; Liu, X. Morphologies and growth mechanisms of α-Al(FeMn)Si in Al–Si–Fe–Mn alloy. *Mater. Lett.* **2013**, *110*, 191–194. [CrossRef]
28. Qi, M.; Kang, Y.; Zhou, B.; Liao, W.; Zhu, G.; Li, Y.; Li, W. A forced convection stirring process for Rheo-HPDC aluminum and magnesium alloys. *J. Mater. Process. Technol.* **2016**, *234*, 353–367. [CrossRef]
29. Zhou, B.; Kang, Y.L.; Zhu, G.M.; Gao, J.Z.; Qi, M.F.; Zhang, H.H. Forced convection rheoforming process for preparation of 7075 aluminum alloy semisolid slurry and its numerical simulation. *Trans. Nonferrous Met. Soc. China* **2014**, *24*, 1109–1116. [CrossRef]
30. Que, Z.P.; Mendis, C.L. Heterogeneous nuclestion and phase transformation of Fe-rich intermetallic compounds in Al-Mg-Si alloys. *J. Alloys Compd.* **2020**, *836*, 155515. [CrossRef]
31. Cao, X.; Campbell, J. The nucleation of Fe-Rich phases on oxide films in Al-11.5Si-0.4Mg cast alloys. *Metall. Mater. Trans. A* **2003**, *34*, 1409–1420. [CrossRef]
32. Miller, D.N.; Lu, L.; Dahle, A.K. The role of oxides in the formation of primary iron intermetallics in an Al-11.6Si-0.37Mg alloy. *Metall. Mater. Trans. B: Process Metall. Mater. Process. Sci.* **2006**, *37*, 873–878. [CrossRef]
33. Zhang, Y.; Jie, J.; Gao, Y.; Lu, Y.; Li, T. Effects of ultrasonic treatment on the formation of iron-containing intermetallic compounds in Al-12%Si-2%Fe alloys. *Intermetallics* **2013**, *42*, 120–125. [CrossRef]
34. Flemings, M.C. *Solidification Process*; Macgrraw-Hill: New York, NY, USA, 1974.

materials

Article

Experimental and Numerical Study on the Influence of Stress Concentration on the Flexural Stability of an Aluminium Hollow Tube

Ganesh Radhakrishnan [1], Daniel Breaz [2,*], Sami Sulaiman Al Khusaibi [1], Amjad Juma Al Subaihi [1], Al Azhar Zahir Al Ismaili [1], AlSalt Malik AlMaani [1] and Kadhavoor R. Karthikeyan [3,*]

1 Mechanical and Industrial Section, Engineering Department, University of Technology and Applied Sciences, Nizwa P.O. Box 477, Oman
2 Department of Mathematics, "1 Decembrie 1918" University of Alba Iulia, 510009 Alba Iulia, Romania
3 Department of Applied Mathematics and Science, National University of Science and Technology, Muscat P.O. Box 620, Oman
* Correspondence: dbreaz@uab.ro (D.B.); karthikeyan@nu.edu.om (K.R.K.)

Abstract: In recent times, particularly in applications used to build various structures for construction purposes or machines, solid sections have been gradually replaced by hollow sections due to their attractive features such as being light weight and having high specific strength. In the present investigation, an attempt was made to investigate, in detail, the flexural capability of aluminium hollow tubes (AHTs) with square cross-sections. The objective of the investigation was to study the influence of stress concentration on the flexural behaviour of the hollow tube. The stress concentration factor considered in this investigation was holes of various cross-sections and quantities. Three-point bending tests with concentrated loads were conducted on specimens of a hollow tube with different stress concentrations such as circular holes, multiple circular holes, square holes and perforations. The load was applied manually during the bending test with appropriate increments. The bending test was carried out on specimens with support spans of 110, 130, 170 and 200 mm. The output measures of the study were maximum bending load, deflection and flexural stiffness. The output measures were analysed in detail in order to recommend the type and nature of stress concentration in a hollow tube applied to structural applications to ensure the safest workability. The flexural stability of the tube was analysed by experimental and numerical procedures, and the results were validated using an analytical approach. It was found that the results of all the approaches complement each other with a low significance of error. AHTs with a circular hole, multiple circular holes and perforations were observed to have better flexural stability than other AHTs such as AHTs with square hole and plain AHTs.

Keywords: hollow tube; flexural strength; stress concentration; flexural stiffness; bending load

Citation: Radhakrishnan, G.; Breaz, D.; Al Khusaibi, S.S.; Al Subaihi, A.J.; Al Ismaili, A.A.Z.; AlMaani, A.M.; Karthikeyan, K.R. Experimental and Numerical Study on the Influence of Stress Concentration on the Flexural Stability of an Aluminium Hollow Tube. *Materials* **2023**, *16*, 1492. https://doi.org/10.3390/ma16041492

Academic Editors: Mostafa Hassani, Hongze Wang and Greta Lindwall

Received: 28 December 2022
Revised: 27 January 2023
Accepted: 7 February 2023
Published: 10 February 2023

1. Introduction

In recent times, most of the engineering applications in the automobile, aerospace, structural member, etc., industries have consistently replaced conventional bulk materials with high-performance and high specific-strength materials such as super alloys and composites and are slowly replacing the solid sections with hollow sections, which not only minimizes the amount of material used in construction but also increases their strength and stability by several times, meeting the desired requirements for which it is intended for. Conventional bulky materials such as steel and iron were replaced by aluminium alloys, magnesium alloys, stainless steel, composite materials, etc., which offer superior specific strength and are economical too. The structural members used in engineering applications are available in many sections, either in the solid or hollow types. Hollow sections have more specific strength than that solid sections. Hollow sections are lighter than solid sections and, therefore, mostly preferred in many structural applications. Square and circular hollow tubes have many applications in the area of multidirectional loading due to their uniform geometry along two or more cross-sectional axes. This, in turn, has uniform strength across the complete structure, which makes them good choices for many structural applications. In spite of heat treatment and super finishing of the materials used in the structural applications, certain machining operations such as drilling, threading, riveting, etc., are mandatory in the case of assembling the structures. These secondary machining operations include the concentration of heavy stress at localized spots around the discontinuities such as holes, notches, etc., that exist in the members for fasteners or any assembly work [1–3]. This type of discontinuity is very difficult to eliminate in the assembly and leads to the concentration of stress at localized spots called stress risers. Stress concentration plays a vital role in affecting the performance of the structure. Stress concentration is the acquisition of a large quantity of stress across a structural member in a particular location due to a sudden change in the geometry or some kind of interruption. Sharp corners, cracks, holes, notches, etc., increase localized stress at the specific location around the interruption, which may lead to failure of the member. The influence of stress concentration need not be the same for solid and hollow sections. The type of stress concentration in the structural member subjected to various loads such as axial, bending and shear, greatly influence the flexural performance [4]. This work was an attempt to investigate, in a detailed manner, the influence of stress concentration in hollow sections subjected to bending load. A significant amount of research has been performed in the area of structural analysis for various sections against various loads including bending; however, flexural stability analyses of hollow sections and comparisons of the results with analytical and numerical analyses to minimize error have not been explored in depth.

A hollow section used for a structure is a type of construction used in many structural applications. It can be circular, square, rectangular, or any other section based on the need. Square and circular hollow tubes have many applications in the area of multidirectional loading due to the uniform geometry along two or more cross-sectional axes. They have uniform strength across the complete structure, which makes them good choices for many structures [5–8]. They also have high resistance to torsion. Zingaila et al. [2] analysed the flexural behaviour of concrete UHPFRC/RC composite members and found that composite beams have enhanced flexural capacity, reduced crack dimensions and increased stiffness compared to that of RC beams. The three-point bending strength of thin silicon dies was evaluated by Tsai et al. [3] for its non-linearity behaviour during bending and concluded, finally, that the correction factor considered in the non-linearity theory highly depends on the deflection, span length, elastic modulus and thickness of the specimens. Wang et al. [5] analysed the tensile and compressive behaviour of carbon fibre-reinforced polyphenylenesulfide composites and found that the strength of the composite material was influenced more by the temperature than the stiffness of the fibre content in the composite. The failure of the composite material was due to the temperature gradient, which was evident from the SEM micrographs. The macro-fracture morphologies from the SEM micrographs illustrated the multiple failure modes due to thermal stress induced in the composite material. Lin et al. [7] compared the flexural behaviour of concrete-filled steel tubular frames with conventional reinforcement concrete structures and analysed various factors that influence the flexural behaviour of these structures. It was concluded in such a way that certain design modifications could be incorporated for the reasonable failure of steel-reinforced concrete members. Tuan et al. [9] have conducted the three-point bending test on concrete-filled steel tubes and observed that the buckling across the tube was delayed due to concrete. In addition, the concrete and steel in the structure complemented each other; therefore, the ductility of concrete was improved by steel, and the compression behaviour of steel was improved by concrete. Aydna et al. [10] investigated the bending and shear performance of the composite fabricated by pouring the waste polymer into the cold-formed I and U profile melds after homogenous pulping. The improved adherence between the steel and polypropylene increased the shear and bending capability. Changing the cross-sectional area in I and U beams under bending influenced the load at yielding, ductility and energy dissipation capacity. The addition of CFRP in I beams significantly increased the bending resistance in the free end region under the shear force. The addition of GFRP bars with better flexibility in I and U beams caused more ductile behaviour than CFRP bars. Zahedi et al. [11] performed their investigation on the flexural behaviour of steel tubes wrapped with carbon-reinforced polymer. Dimensional error was analysed during the study and found that the strength and stiffness were appreciably increased by using FRP laminates around the steel tube. Many research findings are available in this field of hollow tubes, with or without filling, and a considerable amount of scientific concepts and theories have been offered. Some of the important findings from the literature review include that the flexural behaviour of a hollow tube was greatly influenced by the cross-section of the tube, flexural strength of hollow tubes was improved by replacing conventional materials with composite tubes, numerical analysis using ANSYS for flexural testing helps to predict the performance accurately, and the flexural behaviour of a hollow tube using numerical analysis was compared with experimental testing and found to complement each other. The details of the experimentation of the present study are explained in the next section [9,12–16].

2. Experimentation

The flexural behaviour of hollow tubes was investigated. The material used in the investigation was commercially available aluminium hollow tubes of square cross-section with dimensions of 20 × 20 × 1.5. The aluminium used in the hollow tube was Al 6061. The composition and properties of AA 6061 are shown in Tables 1 and 2, respectively.

Table 1. Chemical composition of AA6061.

Chemical Element	% Present
Manganese	0.0–0.15
Iron	0.0–0.7
Magnesium	0.80–1.2
Silicon	0.40–0.8
Copper	0.14–0.4
Zinc	0.0–0.25
Titanium	0.0–0.15
Chromium	0.04–0.35
Others	0.0–0.14
Aluminium	Balance

Table 2. Properties of AA6061.

Property	Value
Mass density	$2.7\ g/cm^3$
Melting Point	652 °C
Thermal expansion	$23.4 \times 10^{-6}/K$
Modulus of Elasticity	71 GPa
Thermal conductivity	168 W/mK
Electrical resistivity	$0.042 \times 10^{-6}\ \Omega m$
Proof stress	245 MPa
Tensile strength	255 MPa
Brinell hardness	94.5 HB

The flexural capability of the aluminium hollow tubes (AHTs) was investigated in detail by conducting three-point flexural tests in a Universal tester machine of 20 kN capacity (Supplier: Gunt, Hamburg, Germany). A concentrated load was applied at the mid span of the specimen, perpendicular to the axis of the tube. The analysis of flexural capability of AHTs was carried out as a plain specimen and specimen with stress risers as well. Holes of different shapes and types were introduced to the AHTs in order to investigate their influence on the flexural capability of the specimen. Five different types of specimens were used in this flexural study. Plain AHTs, AHTs with a through square hole at the midspan on one of the lateral faces with dimension 10 mm, AHTs with through circular hole at the midspan on one of the lateral faces with dimension 5 mm, AHTs with multiple circular through holes at the midspan equally spaced on one of the lateral faces with dimensions 10 mm and 5 mm, and AHTs with perforations of multiple through holes of diameter 10 mm and 5 mm on adjacent lateral faces equally spaced throughout the length. Each test was repeated three times to obtain the results more accurately. The experiment was performed at normal room temperature, and the load applied during the test was manual. The output of the flexural test was taken through a data acquisition system (DAQ), attached to the machine and displayed. The performance measures considered in the flexural test were bending load, deflection and flexural stiffness. The critical bending load capacity and deflection were measured at the midspan of the specimen. The critical bending load was considered an indicator of the maximum resistance offered by the specimen against bending with minimum deflection. The flexural stiffness was calculated using the empirical relation between the critical bending load and the deflection. It was the measure of resistance against deformation, which was greatly influenced by the geometrical properties of the specimen and the loading characteristics. The schematic diagram of flexural test setup is shown in Figure 1a, and the cross-section of the specimen is shown in Figure 1b. The total length of the specimen was 300 mm, and the supports were placed at a distance of 15 mm from the end of the specimen on each side so that the effective length of the specimen under bending was 270 mm. The support rollers were 10 mm in diameter, and the contacts between the supports and the specimen were taken as point contacts with negligible friction. The thickness of all the specimens were 1.5 mm. The roller through which the bending load was applied was also of 10 mm in diameter, and the contact between the loading roller and the loading ram was also ignored during the experimentation. The experimental setup is shown in Figure 2. The same experimentation was also confirmed through a numerical approach using the ANSYS software package. The flexural test setup with the specimen was modelled using the SOLIDWORKS package and imported to ANSYS and analysed. Appropriate boundary conditions and constraints suitable for a three-point bending test were applied as per the standard procedure, and the analysis was carried out. Fine mesh with tetragonal elements of size 1 mm was employed in the finite element study. The load was treated as a concentrated type applied at the midspan of the beam in the transverse direction. The friction between the loading roller and the beam and between the supporting rollers and beam were ignored in the finite element study. This was due to the fact that the bending load was static. The load was applied on a regular and gradual increment basis of approximately 1 kN in each step. The test was repeated twice with the same load in order to ensure consistency of the results. The sensitivity of the element or mesh size was determined based on the literature and the dimensions of the AHT tube used in the study [17]. The material properties used in the experimental approach and the analytical approach were also used for the numerical analysis. Each and every specimen was tested numerically for different support spans under static flexural analysis [8,10,11,18–20]. The experimental bending load was applied to the specimen, and the maximum deflection was noted from the ANSYS results in order to confirm the accuracy of the test output. The model was created using a similar setup to the experimental approach. Bottom rollers were used as supports, whereas the top roller was used to apply the transverse load during the flexural test. The flexural performance of AHTs was validated using an analytical approach with the help of fundamental relations governing the flexural behaviour of a structure.

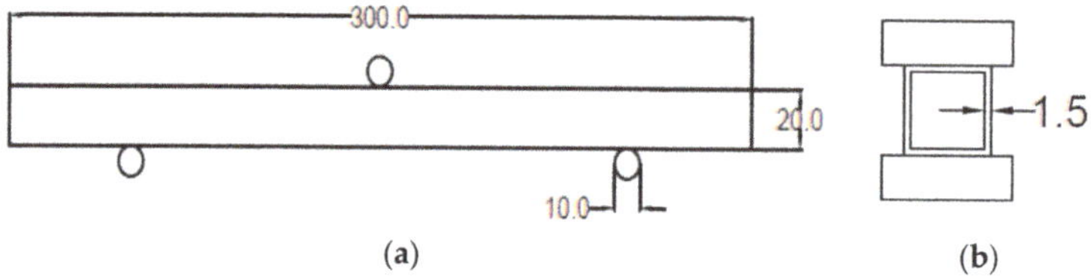

Figure 1. (**a**) Schematic diagram of flexural setup. (**b**) Cross section of AHT.

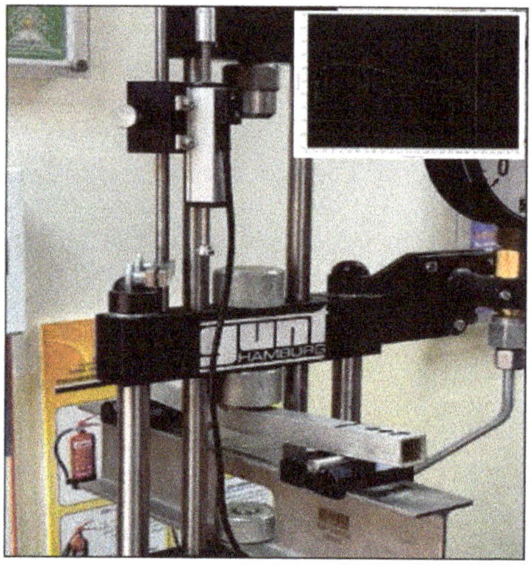

Figure 2. Photograph of experimental setup with a sample output.

3. Results and Discussions

The flexural performance measures such as maximum bending load, maximum deflection and flexural stiffness, which were determined experimentally, are shown in Table 3. From the graphical illustrations shown in Figure 3a–c, it was observed that both the support span and type of discontinuity have significant effects on the flexural stability of AHTs. The support span aspect ratio S/t = 73.33 resulted in better flexural performance compared to that of other aspect ratios, irrespective of the type of discontinuity in the AHT. The lowest bending load capacity was observed for AHTs with square hole, which was attributed to the fact that the sharp corners of the square hole with stress concentration factors of 1.01 increased the concentration of stress around the discontinuity to the maximum level, which in turn increased the severity of failure and, therefore, reduced the bending load capacity. Maximum deflection was observed for the aspect ratio S/t = 133.33, and the least deflection was observed for the aspect ratio S/t = 73.33. The deflection measured at the centre of the specimen followed an increasing trend with respect to an increase in S/t. The longer the support span, the higher the deflection. A larger deflection of a member can result in permanent deflection, cracking and other damage. The larger deflection of one member in a structure may impact its integrity, that of any other member in the structure, or the stability of the entire structure as a whole.

Flexural stiffness was observed to be maximum for AHTs with circular holes and perforations compared to that of AHTs with square holes and plain AHTs as well. This was due to the fact that creating additional holes on the specimen on either side reduced the effect of stress concentration, which in turn increased the stiffness of the specimen. The area moment of inertia of the specimen about minimum cross-sectional area was reduced for the specimens where the stress concentration effect was reduced by additional holes. This in turn increased the resistance to deformability or the flexural stiffness of the specimen under loading. The highest flexural stiffness in the range of 16 to 18 kN/mm was observed for AHTs with square holes. In order to ensure the consistency and accuracy of the results obtained through experimental approach, a numerical analysis using ANSYS software was made with constraints and boundary conditions similar to those of the conditions and assumptions made in the experimental approach. The results of the numerical study are listed in Table 4. The numerical outputs of the flexural test of AHTs are shown in Figure 4a–c. The outputs of the numerical study for deformed specimens with different types of discontinuities and support spans are shown in Figure 5a–e. It was noticed from the illustrations that the bending load capacity was maximum for the support span of S/t = 73.33 and the least for the support span of S/t = 133.33. The numerical results complemented the results obtained through the experimental approach. The influence of support span aspect ratio, S/t, was more on the flexural performance of AHTs rather than the type of discontinuity in the specimen. The numerical results of the flexural test of AHTs revealed that the consistency and accuracy of the experimentation were better and had the least difference in the results. The dimensions of cracks were measured for each failed specimen under flexural loading, tabulated in Table 5. The influence of the type of discontinuity and the support span aspect ratio, S/t, on the crack dimensions are shown in Figure 6. Crack dimensions were observed to be maximum for AHTs with circular and square holes compared to that of AHTs with multiple circular holes and perforations. The support span influenced the crack dimensions significantly. Crack propagation was high for AHTs with circular and square holes. This was due to the fact that the highest stress concentration initiates a crack at an earlier stage and propagates its growth significantly. The crack propagation seems to be faster for specimens with higher concentrations of stress risers. Average crack width seems to be minimum for AHTs with perforations, whereas they are maximum for AHTs with square holes. The lowest crack length of 3.4 mm was observed for AHTs with perforations specimen with aspect ratio, S/t = 113, whereas crack length was a maximum of about 7.5 mm for AHTs with circular holes at aspect ratio S/t = 93 [21,22].

Table 3. Experimental flexural performance measures.

S. No.	Specimen	Stress Concentration Factor (K)	S/t	Maximum Bending Load (kN)	Maximum Deflection (mm)	Flexural Stiffness (kN/mm)
1	Plain	1	73.33	5.759	0.884	6.515
2			93.33	3.624	0.583	6.216
3			113.33	4.12	0.904	4.558
4			133.33	3.275	1.213	2.700
5	Circular hole	1.007	73.33	5.791	0.314	18.443
6			93.33	4.651	0.358	12.992
7			113.33	3.691	0.429	8.604
8			133.33	3.039	0.622	4.886
9	Square hole	1.01	73.33	4.243	0.423	10.031
10			93.33	3.568	0.444	8.036
11			113.33	2.936	0.936	3.137
12			133.33	2.123	0.677	3.136
13	Multiple circular holes	1.01	73.33	4.535	0.288	15.747
14			93.33	3.634	0.317	11.464
15			113.33	2.867	0.454	6.315
16			133.33	2.464	0.639	3.856
17	Perforation	1.014	73.33	4.856	0.362	13.414
18			93.33	4.063	0.647	6.280
19			113.33	3.016	0.779	3.872
20			133.33	2.808	0.634	4.429

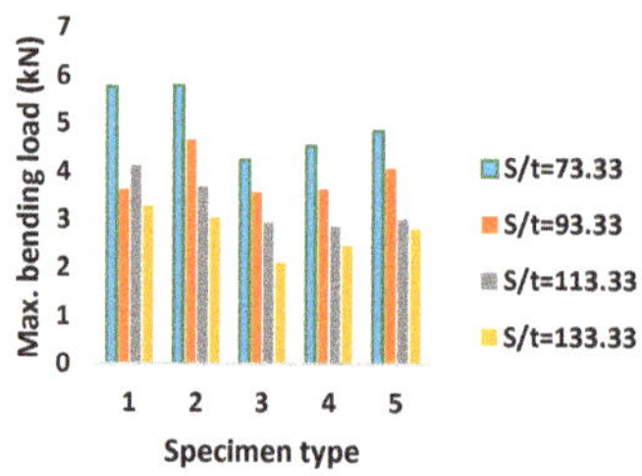

(a)

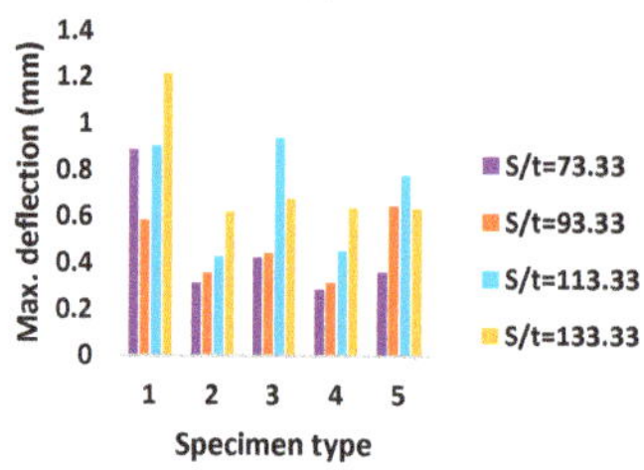

(b)

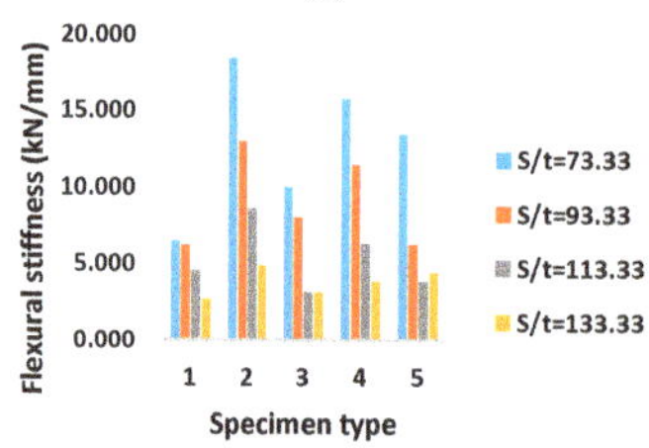

(c)

Figure 3. Experimental flexural performance measures (**a**) Maximum bending load (**b**) maximum deflection and (**c**) Flexural stiffness.

Table 4. Numerical flexural performance measures.

S. No.	Specimen	Stress Concentration Factor (K)	S/t	Maximum Bending Load (kN)	Maximum Deflection (mm)	Flexural Stiffness (kN/mm)
1	AHT with circular hole	1	73.33	5.953	0.465	12.802
2			93.33	5.375	0.865	6.214
3			113.33	2.774	0.8	3.468
4			133.33	1.625	0.763	2.130
5	AHT with square hole	1.007	73.33	4.931	0.388	12.709
6			93.33	4.087	0.663	6.164
7			113.33	2.3	0.668	3.443
8			133.33	1.503	0.711	2.114
9	AHT with multiple circular holes	1.01	73.33	4.566	0.362	12.613
10			93.33	2.429	0.397	6.118
11			113.33	2.047	0.599	3.417
12			133.33	1.173	0.559	2.098
13	AHT with perforation	1.01	73.33	5.513	0.437	12.616
14			93.33	2.613	0.427	6.119
15			113.33	1.756	0.514	3.416
16			133.33	1.368	0.652	2.098
17	AHT with circular hole	1.014	73.33	4.855	0.683	7.108
18			93.33	4.942	0.828	5.969
19			113.33	3.216	0.965	3.333
20			133.33	1.865	0.911	2.047

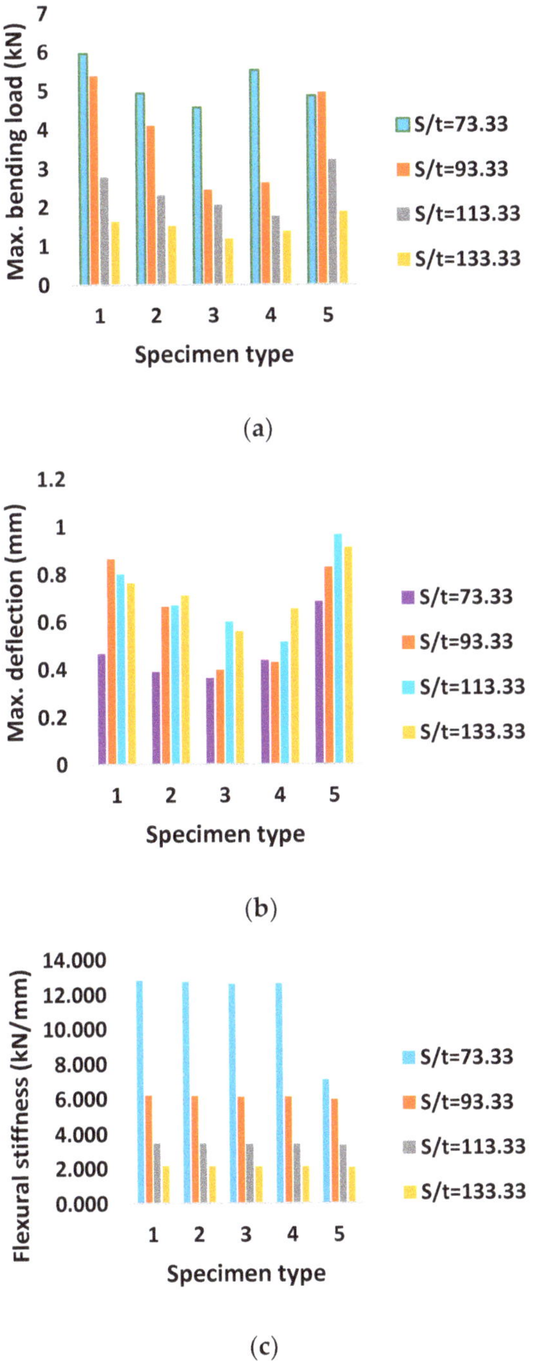

Figure 4. Numerical flexural performance measures (**a**) Maximum bending load (**b**) maximum deflection and (**c**) Flexural stiffness.

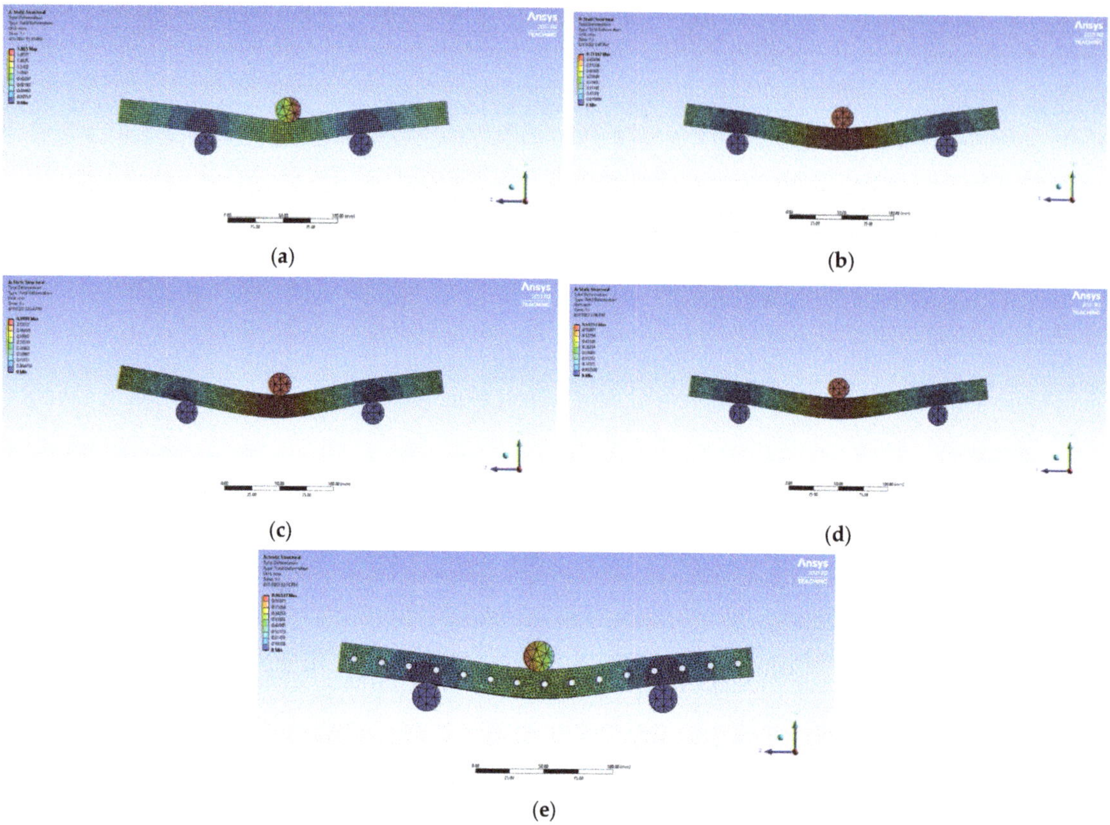

Figure 5. Sample outputs of numerical flexural performance at maximum deflection (**a**) plain AHT (**b**) AHT with circular hole (**c**) AHT with square hole (**d**) AHT with multiple circular holes and (**e**) AHT with perforations.

Table 5. Crack dimensions of flexural failed specimens.

S. No.	Specimen	Aspect Ratio (S/t)	Stress Concentration Factor (K)	Crack Width (mm)	Crack Length (mm)
1	AHT with circular hole	73	1.007	1.1	6.9
2		93		1.8	7.5
3		113		1.3	6.85
4		133		1	6.8
5	AHT with square hole	73	1.01	1.8	4.5
6		93		2.1	4.7
7		113		2.83	4.25
8		133		2.45	4.7
9	AHT with multiple circular holes	73	1.01	1.66	4.85
10		93		1.32	4.25
11		113		1.73	4.6
12		133		0.95	3.8
13	AHT with perforation	73	1.014	1.1	4.56
14		93		0.8	4.75
15		113		1.05	3.4
16		133		1.2	4.78

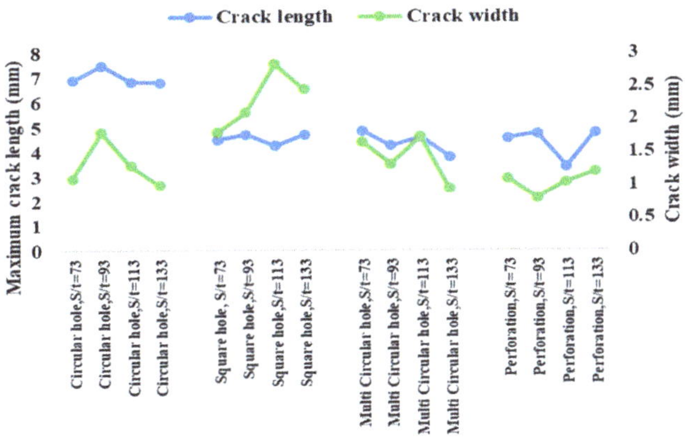

Figure 6. Crack dimensions of flexural failed specimens.

4. Conclusions

An attempt was made successfully to study the flexural stability of aluminium hollow tube (AHTs) with different discontinuities in the form of holes of different geometries such as circular holes, square holes, multiple circular holes and perforations. The aspect ratio support span to thickness of the tube (S/t) was considered as the process parameter during the flexural test of AHTs. The study was performed in two aspects, experimental and numerical, and fruitful outcomes and conclusions were drawn from the study and listed below.

(i) The support span aspect ratio (S/t) significantly influenced the flexural behaviour of plain AHTs and AHTs with holes of different quantities and geometries.

(ii) Following the support span aspect ratio, the type of discontinuity in terms of shape, size and quantity significantly influenced the flexural stability of AHTs. This may be due to the fact that these discontinuities acted as stress risers and influenced the flexural capacity of the structure to a larger extent.

(iii) AHTs with a circular hole, multiple circular holes and perforations were observed to have better flexural stability than that of other AHTs such as AHTs with square holes and plain AHTs. This may be attributed to the fact that the sharp corners in the holes are the sources of crack initiation and propagation, which in turn lead to the failure of the specimen under loading.

(iv) Reducing the effect of stress concentration abruptly increased the flexural behaviour of AHTs by offering better flexural resistance.

(v) The results obtained through experimental and numerical approaches complemented each other with utmost accuracy and consistency. The assumptions and constraints made during the analysis were very close and complemented each other.

Author Contributions: Conceptualization, G.R., D.B., S.S.A.K., A.J.A.S., A.A.Z.A.I., A.M.A., K.R.K.; methodology, G.R., D.B., S.S.A.K., A.J.A.S., A.A.Z.A.I., A.M.A., K.R.K.; software, G.R., D.B., S.S.A.K., A.J.A.S., A.A.Z.A.I., A.M.A., K.R.K.; validation, G.R., D.B., K.R.K.; formal analysis, G.R., D.B., S.S.A.K., A.J.A.S., A.A.Z.A.I., A.M.A., K.R.K.; investigation, G.R., D.B., S.S.A.K., A.J.A.S., A.A.Z.A.I., A.M.A., K.R.K.; resources, G.R., D.B., S.S.A.K., A.J.A.S., A.A.Z.A.I., A.M.A., K.R.K.; data curation, G.R., D.B., S.S.A.K., A.J.A.S., A.A.Z.A.I., A.M.A., K.R.K.; writing—original draft preparation, G.R., D.B., S.S.A.K., A.J.A.S., A.A.Z.A.I., A.M.A., K.R.K.; writing—review and editing, G.R., D.B., S.S.A.K., A.J.A.S., A.A.Z.A.I., A.M.A., K.R.K.; visualization, G.R., D.B., S.S.A.K., A.J.A.S., A.A.Z.A.I., A.M.A., K.R.K.; supervision, G.R., D.B.; project administration, G.R., D.B.; funding acquisition to pay for the APC., D.B. All authors have read and agreed to the published version of the manuscript.

Funding: This research received no external funding.

Institutional Review Board Statement: Not applicable.

Informed Consent Statement: Not applicable.

Data Availability Statement: The data that support the findings of this study are available within the article.

Conflicts of Interest: The authors declare no conflict of interest.

References

1. Scărlătescu, D.D.; Modrea, A.; Stanciua, M.D. Three-point bend test to determine the mechanical behavior of the tubes used in water supply networks. *Procedia Manuf.* **2019**, *32*, 179–186. [CrossRef]
2. Zingaila, T.; Augonis, M.; Arruda, M.R.T.; Serelis, E.; Kelpsa, S. Experimental and numerical analysis of flexural concrete-UHPFRC/RC composite members. *Mechanika* **2017**, *23*, 182–189. [CrossRef]
3. Tsai, M.-T.; Huang, P.S.; Yeh, J.H.; Liu, H.Y.; Chao, Y.C.; Tsai, F.; Chen, D.L.; Shih, M.K.; Tarng, D. Evaluation of three-point bending strength of thin silicon die with a consideration of geometric nonlinearity. *IEEE Trans. Device Mater. Reliab.* **2019**, *19*, 615–621. [CrossRef]
4. Buyukkaragoz, A.; Kalkan, I.; Leec, J.H. A numerical study of the flexural behavior of concrete beams reinforced with afrp bars. *Strength Mater.* **2013**, *45*, 716–729. [CrossRef]
5. Wang, S.; Zhang, J.; Zhou, Z.; Fang, G.; Wang, Y. Compressive and flexural behavior of carbon fiber-reinforced PPS composites at elevated temperature. *Mech. Adv. Mater. Struct.* **2020**, *27*, 286–294. [CrossRef]
6. Shallal, M.A. Flexural behavior of concrete filled steel tubular beam. In Proceedings of the 2018 International Conference on Advance of Sustainable Engineering and Its Application (ICASEA), Wasit, Iraq, 14–15 March 2018; pp. 153–158. [CrossRef]
7. Liu, Y.-B.; Cui, P.-P.; Chen, F. On factors behind the reasonable failure mode of concrete-filled circular steel tubular composite frame. *Adv. Mater. Sci. Eng.* **2021**, *2021*, 3027640. [CrossRef]
8. Al Zand, A.W.; Ali, M.M.; Al-Ameri, R.; Badaruzzaman, W.H.W.; Tawfeeq, W.M.; Hosseinpour, E.; Yaseen, Z.M. Flexural Strength of Internally Stiffened Tubular Steel Beam Filled with Recycled Concrete Materials. *Materials* **2021**, *14*, 6334. [CrossRef]
9. Tuan, C.Y. Flexural behavior of non-posttensioned and posttension concrete filled circular steel tubes. *J. Struct. Eng.* **2008**, *134*, 1057–1060. [CrossRef]
10. Aydna, A.C.; Bayraka, B.; Maali, B.; Mete, E.; Cebi, K.; Kilic, M. The shear and flexural behavior of cold-formed steel composite I and U beams. *Sci. Iran. A* **2020**, *28*, 2119–2132. [CrossRef]
11. Zahedi, L.; Oskouei, A.V.; Rajaee, S. A Numerical Study on Steel Tubes Wrapped with CFRP Laminates: Carbon Fiber Reinforced Polymers Applications on Offshore Marine Structures. *Int. J. Constr. Environ.* **2013**, *3*, 75–86. [CrossRef]
12. Ibrahim, A.M.; Salman, W.D.; Bahlol, F.M. Flexural behavior of concrete composite beams with row steel tube section and different shear connectors. *Tikrit J. Eng. Sci.* **2019**, *26*, 51–61. [CrossRef]
13. Marcadon, V.; Kurch, S. Roles of mechanical heterogeneities and damage on the overall mechanical behaviour of hollow-tube stackings. *Eng. Procedia* **2011**, *10*, 2815–2820. [CrossRef]
14. Rong, B.; Guo, Y.; Li, Z. Study on the stability behavior of 7A04-T6 aluminum alloy square and rectangular hollow section columns under axial compression. *Build. Eng.* **2022**, *45*, 103652. [CrossRef]
15. Awad, Y.D.; Al-Ahmed, A.H.A. Performance of hollow core concrete slab reinforced by embedded steel tubes. *Assoc. Arab. Univ. J. Eng. Sci.* **2019**, *26*, 17–21. [CrossRef]
16. Zhou, J.; Wen, Z.; Mao, W.; Zhong, C.; Wang, K.; Zhou, C. Full-Scale Model Experimental Study of the Flexural Behavior of Hollow Slabs Strengthened by UHPC. *Adv. Mater. Sci. Eng.* **2021**, *2021*, 5581022. [CrossRef]
17. Nassiraei, H.; Rezadoost, P. Stress concentration factors in tubular T joints reinforced with external ring under in-plane bending moment. *Ocean. Eng.* **2022**, *266*, 112551. [CrossRef]
18. Johannesssen, H.; Johannessen, O.H.; Costas, M.; Clausen, A.H.; Sønstabø, J.K. Experimental and numerical study of notched SHS made of different S355 steels. *Constr. Steel Res.* **2021**, *182*, 106673. [CrossRef]
19. Barbosa de Oliveira, M.A.; Rodrigues da Cunha, R.; de Souza Picanço, M.; Carvalho de Oliveira, D.R.; Leal Soares Ramos, E.M.; Pereira-da-Silva, M. Analysis of the influence of test method and properties of steel fiber addition on concrete under the three-point flexural tensile. Ingeniare. *Rev. Chil. Ing.* **2020**, *28*, 373–382. [CrossRef]
20. Diab, H.M.; Abdelaleem, T.; Rashwan, M.M.M. Flexural behavior of RC continuous t-beams reinforced with hybrid CFRP/ steel bars: Experimental and numerical study". *J. Eng. Sci. Assiut Univ. Fac. Eng.* **2021**, *49*, 215–247. [CrossRef]
21. Ganesh, R.; Saravanan, M. Effect of fiber orientation on mechanical behavior of glass fiber reinforced polyethylene terephthalate foam sandwich composite. *Mater. Today Proc.* **2022**, *62*, 624–628. [CrossRef]
22. Radhakrishnan, G.; Hattali, A.; Haitham, A.; Al Yahyai, A.M.; Al Riyami, A.M.; Al Hadhrami, A.M. Experimental study on the effect of aspect ratio on flexural behavior of Aluminium Sandwich Composite. *Eng. Technol. J.* **2022**, *40*, 990–995. [CrossRef]

materials

Article

Microstructure and Mechanical Properties of Hypereutectic Al-High Si Alloys up to 70 wt.% Si-Content Produced from Pre-Alloyed and Blended Powder via Laser Powder Bed Fusion

Jan Henning Risse [1], Matthias Trempa [1,*], Florian Huber [2], Heinz Werner Höppel [3], Dominic Bartels [2], Michael Schmidt [2], Christian Reimann [1] and Jochen Friedrich [1]

1 Fraunhofer IISB, Schottkystrasse 10, 91058 Erlangen, Germany
2 Department of Mechanical Engineering, Institute of Photonic Technologies, Friedrich-Alexander-University of Erlangen-Nuremberg, Konrad-Zuse-Straße 3/5, 91052 Erlangen, Germany
3 Department Material Science and Engineering, Institute I: General Materials Properties, Friedrich-Alexander-University of Erlangen-Nuremberg, Martensstr. 5, 91058 Erlangen, Germany
* Correspondence: matthias.trempa@iisb.fraunhofer.de

Abstract: Hypereutectic Al-high Si alloys are of immense interest for applications in the automotive, space or electronic industries, due their low weight, low thermal expansion, and excellent mechanical and tribological properties. Additionally, their production by laser powder bed fusion (LPBF) technology provides high flexibility in geometrical design and alloy composition. Since, most of the alloy properties could be improved by increasing the Si content, there is much interest in discovering the maximum that could be realized in LBPF Al-high Si alloys, without the appearance of any material failure. For this reason, in this work the production of Al-high Si alloys with extremely high silicon content of up to 70 wt.% was fundamentally investigated with respect to microstructure and mechanical properties. Highly dense (99.3%) and crack-free AlSi50 samples ($5 \times 5 \times 5$ mm^3), with excellent hardness (225 HV5) and compressive strength (742 MPa), were successfully produced. Further, for the first time, AlSi70 LBPF samples of high density (98.8%) without cracks were demonstrated, using moderate scanning velocities. Simultaneously, the hardness and the compressive strength in the AlSi70 alloys were significantly improved to 350 HV5 and 935 MPa, as a result of the formation of a continuous Si network in the microstructure of the alloy. With respect to the powder source, it was found that the application of powder blends resulted in similar alloy properties as if pre-alloyed powders were used, enabling higher flexibility in prospective application-oriented alloy development.

Keywords: additive manufacturing; laser powder bed fusion; hypereutectic Al-high Si alloys; in-situ alloying; microstructure; mechanical properties

Citation: Risse, J.H.; Trempa, M.; Huber, F.; Höppel, H.W.; Bartels, D.; Schmidt, M.; Reimann, C.; Friedrich, J. Microstructure and Mechanical Properties of Hypereutectic Al-High Si Alloys up to 70 wt.% Si-Content Produced from Pre-Alloyed and Blended Powder via Laser Powder Bed Fusion. *Materials* **2023**, *16*, 657. https://doi.org/10.3390/ma16020657

Academic Editors: Mostafa Hassani, Hongze Wang and Greta Lindwall

Received: 17 November 2022
Revised: 13 December 2022
Accepted: 21 December 2022
Published: 10 January 2023

1. Introduction

Hypereutectic Al-high Si alloys, especially with Si content > 30 wt.%, are characterized by preeminent properties, like low density, low thermal expansion, high hardness and stiffness, and excellent wear resistance, which enable numerous applications in the automotive, space and electronics industries [1,2]. All the alloy properties mentioned improve with increasing Si content, hence, there is high industrial interest regarding the commercial availability of AlSi-alloys with Si-content $\geq$ 50 wt.%. However, during the conventional casting process of such Al-high Si alloys, a coarse and highly brittle primary Si phase is formed, leading to crack formation in the casted parts and reducing their performance [3,4]. The Fabrication via rapid solidification technologies, like spray deposition or cold/hot pressing, can prevent the extensive coarsening of this Si phase and enable the production of alloys up to 90 wt.% silicon [5–7]. However, these production methods are limited to rather simple geometries and often require some post-treatment relating to final densification or geometrical finishing.

A promising fabrication technique, which mostly overcomes these limitations, is additive manufacturing (AM), especially the powder bed fusion of metal with laser beam (PBF–LB/M or LPBF) technology, in which complex and dense components are built up via printing layer by layer [8,9]. This technique has already been used to principally investigate Al-high Si alloys with a silicon content up to 50 wt.% (see e.g., [4,10]). A comparison to casted AlSi50 samples, conducted by Jia et al. [4], showed a refinement of the primary Si, from platelike particles, with sizes >100 μm, to polygonal-shaped particles, with sizes <6 μm, using PBF–LB/M, which led to an increase of the compressive strength by almost 50%, to over 650 MPa. Kang et al. [10] demonstrated AlSi50 samples with a maximal hardness of 188 HV0.3, which is approximately 50% higher in comparison to the well-established AM-alloy AlSi10Mg [11]. Despite these promising singular results, knowledge about the microstructure and its correlation to the mechanical properties of these Al-high Si AM-alloys is still limited. So, the findings in the literature about processibility via PBF–LB/M, and the material properties, are not consistent. On the one hand, several researchers reported cracks in AlSi40 [12–14] and AlSi50 samples [15]. On the other hand, other findings did not mention any crack formation [4,10,16]. Consequently, further research is needed to clarify the cracking issue. Additionally, the reported density values of AlSi50 samples are widely scattered, from ~98% [10] to ~99.9% [15], so there is a need to clarify if this difference is related to intrinsic material properties, processing or accuracy in characterization. Further, the production of Al-high Si alloys with Si contents higher than 50 wt.% have not yet been demonstrated, which might be correlated with the cracking issue mentioned, as well as with a lack of standardized AM powder sources.

Therefore, the aims of this work were the following: (1) to extend knowledge about the microstructure and its correlation to the mechanical properties in AM Al-high Si alloys and (2) to prove the feasibility of AM Al-high Si alloys with even higher Si content. A detailed investigation into the correlation of process parameters, especially scan velocity, microstructure, and crack formation, as well as mechanical properties in PBF–LB/M AlSi50 alloys was carried out, followed by extension to an even higher Si content of 70 wt.%. Parallel to these activities, due to the lack of pre-alloyed AlSi70 powder, the effect of the use of powder blends, which were already successfully tested for AlSi40 [13] and AlSi50 [10,15] alloys, was investigated.

2. Materials and Methods

2.1. Materials

To investigate the effect of the powder sources on the resulting alloy properties, three different powder types, namely, a pre-alloyed AlSi50 powder, an Al powder, and a Si powder, were used. The particle size distributions, measured by laser diffraction (particle analyzer Beckman-Coulter LS13320) according to ISO 13320-1, and SEM-images from single powder particles are shown in Figure 1. The pre-alloyed AlSi50 powder (gas atomized by NANOVAL GmbH & Co. KG, Berlin, Germany) had a mainly spherical shape with some satellites on the particle surfaces and a skewed particle size distribution, with a medium diameter (D50-value) of 40 μm. The spherical Al99.7 powder (gas atomized by TLS Technik GmbH & Co. Spezialpulver KG, Niedernberg, Germany) and the edged Si powder (Silgrain® from Elkem ASA, Oslo, Norway) had Gaussian particle size distributions with D50-values of 44 μm and 49 μm, respectively.

On the one hand, the pre-alloyed AlSi50 powder (p.) was compared to the more flexible and cheaper approach of using powder blends made of elementary that were only partially gas-atomized (Al + Si). Furthermore, it was also used in blends with silicon (p. + Si) to achieve a higher silicon content of 70 wt.% and to evaluate the different powder blends in its performances. All powders/powder blends were dried at 110 °C for 15 h under vacuum atmosphere and the powder blends were mixed for 1 h in a tumbling mixer after drying before being further processed.

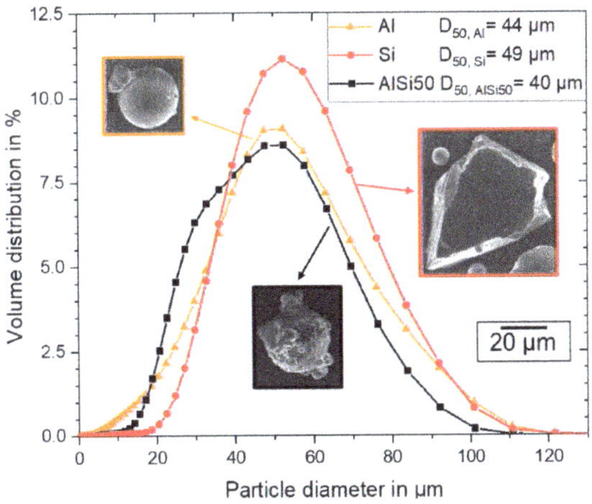

Figure 1. Particle size distribution and SEM images of the elemental Al and Si powders and the pre-alloyed AlSi50 powder measured via laser diffraction.

2.2. Sample Fabrication via PBF–LB/M, Preparation, and Characterization

The samples were produced using a mini PBF–LB/M machine from Aconity GmbH, equipped with a single mode fiber laser with up to 1 kW power and a wavelength of 1080 nm. During the fabrication process the building chamber was ventilated with argon gas to keeping the residual oxygen content below 100 ppm. The build space was downsized to a 55 mm diameter platform made of a sandblasted aluminum base plate, with a thickness of 5 mm.

Based on a preceding process parameter optimization, the laser power, layer thickness, hatch distance, and laser spot size were fixed to 350 W, 50 µm, 120 µm and 72 µm, respectively. Only the influence of the scan velocity was investigated and is discussed in the present article. The laser beam scanned the layer's geometry in a meandering pattern, which was rotated by 67° in the subsequent layer. The microstructure was examined in $5 \times 5 \times 5$ mm^3 cubes, which were ground and polished up to a final polishing step with 0.25 µm fumed silica. To determine the porosity, light microscope images of longitudinal cuts were analyzed using a MATLAB script, which converted the image into a binary image and calculated the porosity, as well as the size, of the single pores. With this measurement procedure the smallest resolvable pore diameter was approximately 1 µm, which is significantly smaller than the 20–30 µm which is typically reached by computer topography (CT). The longitudinal cuts of almost all samples built with scan velocities of 1000, 1400, 1600, 2000 and 2400 mm/s (the last only with AlSi50 p.) were inspected via SEM (JEOL 6510, equipped with a secondary electron detector) to investigate the features of the microstructure. Therefore, the cuts were etched with deionized water and HF and metallized with an approximately 3 nm platinum layer. For each sample, nine micrographs were taken and analyzed via ImageJ regarding the primary Si particle sizes. The fractions of the fine and coarse areas were determined, based on light microscopic images.

The hardness was measured via Vickers hardness testing (type KBW 10-V) with a test force of 49.03 N (HV5) at the same longitudinal cuts as the microstructural analysis. The data, shown in Section 3, consisted of four hardness indentations along the vertical center line of the samples. In addition, uniaxial compression tests were carried out in small cylinders with the of dimensions Ø 5 mm × 7 mm (mechanically machined from built samples with Ø 7 mm × 15 mm) at a compression speed of 10^{-3} s^{-1} using an Instron 4505 equipped with a Hegewald and Peschke control system. For each combination of powder variant and scan velocity the characteristic parameters were determined for four samples. The Young's modulus was determined via the dynamic resonance frequency method on

Ø 5 mm × 50 mm round bars made of AlSi50 p. and AlSi70 Al + Si, which were built with a scan velocity of 1000 mm/s.

3. Results and Discussion

3.1. Porosity and Cracks

In Figure 2a, the measured porosities of all samples are shown in their dependence on scan velocity, powder type and Si content. Herein, one data point represented one sample. The measured values were between 0.4% and ~3% considering all variants. To verify these results, a reproducibility test with eight identical AlSi50 p. samples built with 1600 mm/s within one run was carried out, resulting in an average porosity of 0.54% and a doubled standard deviation of 0.22%. Since the scattering between the samples was much lower than the variations observed in dependence on scan velocity and the different material types, the latter ones could be identified as real effects and not irregularities from the PBF–LB/M process itself.

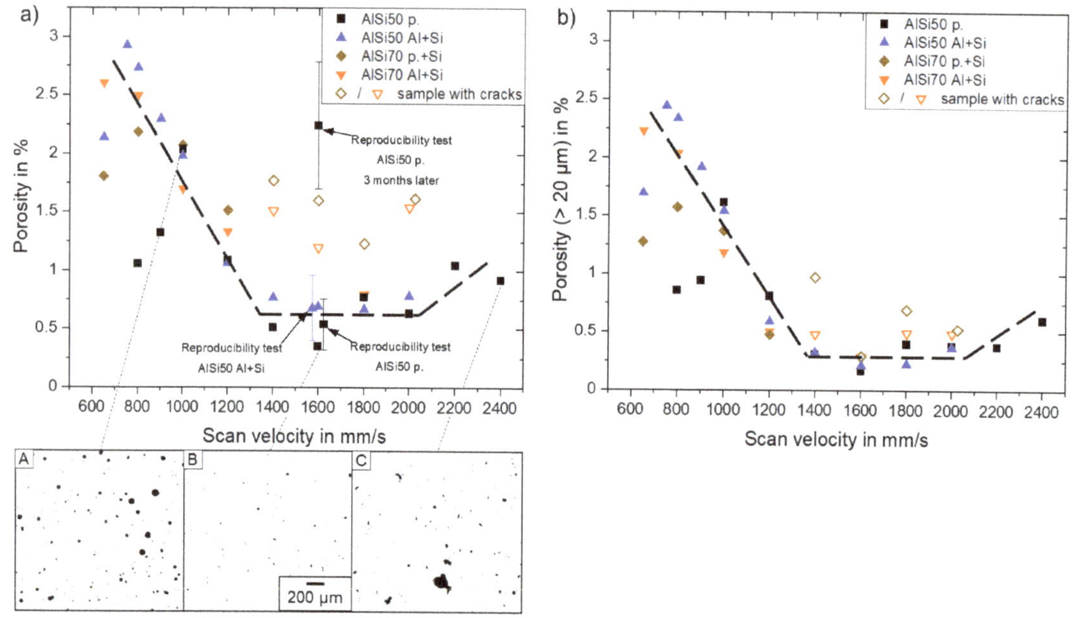

Figure 2. (**a**) Porosity vs. scan velocity for AlSi50 and AlSi70 samples built from different powder types. Binary images of AlSi50 p. samples built with 1000 mm/s (**A**), 1600 mm/s (**B**) and 2400 mm/s (**C**). (**b**) Same as (**a**), but only pores with diameter >20 µm were considered.

The pre-alloyed AlSi50 p., which represents the most conventionally used powder type option, showed its porosity minimum at ~0.5% between 1400 mm/s and 2000 mm/s. In this scan velocity range, only small pores occurred in the polished cuts, as visible in binary image B. One possible cause of these pores could be hydrogen, which is generated through the interaction between the laser beam and residual moisture on the particles' surfaces [17]. With decreasing scan velocity, the porosity rises due to the formation of bigger keyhole pores resulting from a too high energy input [18]. These pores had a spherical shape, as can be observed in binary image A. Likewise, the porosity also increased at scan velocities ≥2200 mm/s, but the geometry of the occurring pores changed from a spherical to an irregular shape (compare binary image C). These pores, lacked fusion defects, and were formed due to a non-sufficient energy input resulting in incompletely melted zones [19]. A similar dependency of the porosity on the scan velocity was reported by Mueller et al. [12] for pre-alloyed AlSi40 with a minimum porosity of approximately 1% at 1350 mm/s.

In contrast to the homogeneous powder properties of the AlSi50 p., the elemental Al and Si powders were differentiated in morphology and size, which could lead to inferior part quality when using the blended AlSi50 Al + Si powder. However, the AlSi50 Al + Si showed its porosity minimum at ~0.7% between 1400 mm/s and 2000 mm/s, which was very close to the minimum of the AlSi50 p. powder, and showed increased porosities at slower scan velocities. The porosity determined by the reproducibility test averaged out at 0.68% with a double standard deviation of 0.28%. Consequently, the two different powder approaches can be seen as equivalent in relation to porosity and, so far, the usage of the blended powder resulted in no disadvantage. This was in good agreement with the results of Garrard et al. [13], who showed similar porosities of AlSi40 samples made of pre-alloyed and blended powders.

Furthermore, the measured porosities were in the range of published values for AlSi50, lying between <2% [10] and approximately 0.1% [15]. Nevertheless, the wide scattering within the same material could either be caused by the process parameters or by the porosity measurement itself, because the resolution of the technique used determined the amount of detectable pores Alongside the optical measurement at cuts, nondestructive computer tomography (CT) is a common method used in additive manufacturing, but is rather limited in resolutions up to ~20 µm [20]. To mimic the results which would be obtained from a CT measurement, Figure 2b shows the porosity of the same samples, but considering only pores larger than 20 µm in diameter. It is obvious, that all porosity values decreased. However, while at slower scan velocities the porosities did not change very much due to the presence of large keyhole pores (compare image A in Figure 2), the porosities of both AlSi powder material variants, between 1400 mm/s and 2000 mm/s, decreased to about 0.3%.

The similarity in porosity, regardless of which powder type was used, appeared also in the case of the AlSi70 variants, namely AlSi70 p. + Si and AlSi70 Al + Si. Compared to the AlSi50 samples, the porosity increased from ~0.7% to approximately 1.5% between 1200 mm/s and 2000 mm/s. However, the "simulated" CT measurement showed porosities of the AlSi70 samples of a similar level as the AlSi50 samples. In consequence, the AlSi70 samples, in particular, contained a conspicuous quantity of small pores, which would not be detected in a CT measurement. As the AlSi70 samples were built three months later than the AlSi50 samples, an ageing of the powders and an absorption of moisture was assumed to be a possible explanation for the increase in the number of fine pores. A second reproducibility test with AlSi50 p. powder confirmed this hypothesis. An extreme increase of fine pores <20 µm (approximately 2000%) resulted in a porosity of >2% (see Figure 2a) in this test. Hardness measurements of the old and new samples built within the reproducibility tests showed that the fine pores did not influence the hardness. Probably, these pores were not crucial for the mechanical performance, but maybe for thermo–physical or electrical properties of the alloys. This must be considered during material development when carrying out sample analysis by means of the CT method.

The occurrence of cracks in Al-high Si material is attributed to its low fracture toughness in combination with high thermal stresses induced by steep thermal gradients, high solidification rates and mismatch in thermal expansion between the AlSi material and the base plate (commonly made of Al) used [15]. The influence of process parameters on the thermal gradient and the solidification rate can be described by the energy density $E = P/(v*d*h)$ [J/mm^3], which is a function of laser power P, scanning speed v, hatch distance h and layer thickness d. Cracking in AlSi40 samples was reported for energy densities <50 J/mm^3 [12] or <66 J/mm^3 [13], respectively. In the case of AlSi50 material, Hanemann et al. [15] observed cracks for energies $\leq$111 J/mm^3, while crack-free samples were obtained by Kang et al. [10] only at even significantly higher energies $\geq$231 J/mm^3. Hence, higher energy densities are supposed to lead to reduced thermal gradients and solidification rates and, therefore, should counteract crack formation in Al-high Si alloys.

However, all produced AlSi50 samples in this work (energy range was 24–97 J/mm^3) showed no cracks in volume, independent of the powder source. This contrasted with the earlier findings described above and pointed out that energy density is a not sufficient parameter to explain the cracking behavior of these alloys. Nevertheless, the results showed that production of crack-free AlSi50 samples, with low porosities, was possible, even at lower energy densities.

In the case of higher Si content, cracks occurred in the AlSi70 samples built with scan velocities of ≥1400 mm/s. They were probably caused by the higher mismatch of thermal expansion with the Al substrate, in combination with the high thermal stresses induced by high scanning velocities. This trend of higher crack sensibility at high scan velocities was also observed for AlSi40 [12,13]. A common counter measure in the literature is a heated baseplate to reduce the thermal stresses [21], but the PBF–LB/M machine used here was not equipped with such a heater system, so this was not further investigated. Nevertheless, it could be successfully demonstrated that crack-free AlSi70 samples with porosities down to 1.25% (1 µm pore resolution) or ~0.5% (CT resolution) could be produced by the PBF–LB/M method using moderate scan velocities ≤1200 mm/s.

3.2. Microstructure

Resulting from the layer-by-layer buildup, additively manufactured, Al-high Si alloys develop a characteristic microstructure, which consists of two different characteristic areas [12,15,16], as shown in Figure 3. On the one hand, in the center of the melting tracks, fine grains occurred, consisting of primary Si crystals. This region is referred to as the "fine area". On the other hand, in the overlapping zones and the periphery of the melting tracks coarser primary Si crystals formed, referred to as the "coarse area". The different grain structure appearance might be correlated to different thermal histories. For example, the temperature distribution and, in consequence, the cooling rate is rather non-uniform inside the melting track [22]. The highest solidification rate occurs in the center of the melt pool, resulting in fine grains. Towards the periphery, the solidification rate continuously decreases, which leads to larger gains. The fine-grained center regions of the melting tracks are typically melted once, while the overlapping coarse-grained regions are melted twice. During the second melting, the formerly built Si phases can partly remain un-melted, so they can act as nucleation sites for new larger Si phases growing during the second solidification [4].

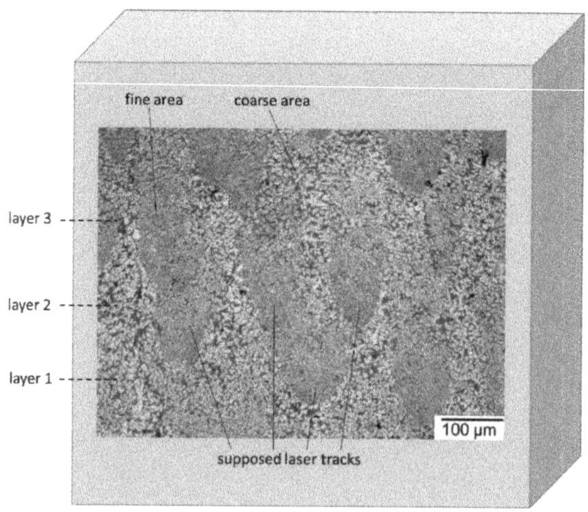

Figure 3. Light microscope image of an etched longitudinal cut of an AlSi50 sample with indicated possible laser tracks and layers.

Figure 4 shows the microstructure of the transition region between fine- and coarse-grained areas for all considered powder variants. In the fine area of the AlSi50 p. sample, the primary Si particles had a branched or blocky morphology with rounded edges, while they were embedded in a single-phase matrix, which was probably a super saturated solid solution of aluminum [4,16]. The morphology of the primary Si in the coarse area changed into polygonal particles with partly sharp edges, which were surrounded by a eutectic matrix with eutectic Si and an Al solid solution. At the transition between these areas the particles were sometimes connected with each other, which was the biggest difference in comparison to the AlSi50 Al + Si, where the areas were not connected via the particles. Furthermore, the particles' edges of the AlSi50 Al + Si were more rounded, and the matrix seemed to contain more eutectic Si, which could be related to impurities acting as nuclei for the Si.

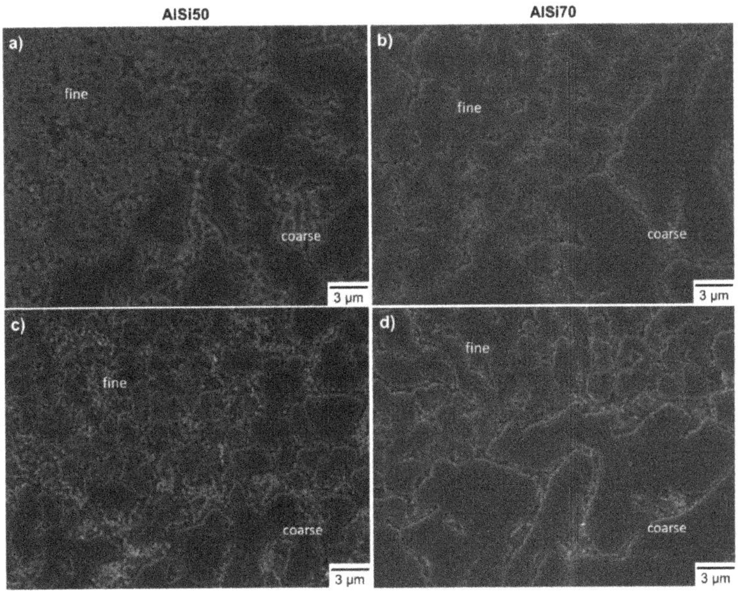

Figure 4. SEM micrographs of the microstructure of: (**a**) AlSi50 p.; (**c**) AlSi50 Al + Si; (**b**) AlSi70 p. + Si and (**d**) AlSi70 Al + Si, recorded in the transition of the fine and coarse area. All samples were built with 1400 mm/s.

Within the AlSi70 samples, the roundly edged primary Si particles built a network in both the fine and the coarse areas. Again, the contiguity in the transition area was more developed in the case of the AlSi70 p. + Si than for that of AlSi70 Al + Si. In analogy to the AlSi50 samples, the interspace of the Si-network was composed of eutectic Si and Al solid solution.

Thus, it can be stated, that the main differences between the microstructures of the four alloy variants were the strongly developed network of primary Si in the AlSi70 samples, compared to the mostly separated Si-phases in the AlSi50 samples, and the higher contiguity in the samples produced with the pre-alloyed powders, AlSi50 p. and AlSi70 p. + Si. The latter could be caused by different interaction behaviors of the laser beam with the powders, e.g., in relation to different laser absorption ratios [23]. Due to the high reflectivity of the Al powder, less energy could be absorbed by the blended powder beds, which might lead to a narrower melt pool. Consequently, the overlapping of the melting tracks, as well as the contiguity of the Si phases in those regions, could be reduced.

A characteristic parameter to describe the microstructure of AlSi alloys is the size of the primary Si phases formed during the fast cooling of the melt. While the particle size could not be determined in the AlSi70 samples, due to the high contiguity of the Si phases, it was measured in the AlSi50 alloys. The results are shown in Figure 5 in dependence on the scan velocity and the powder type.

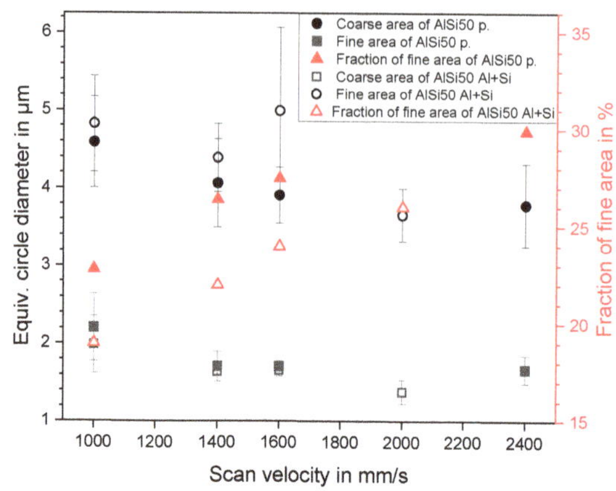

Figure 5. The equivalent circle diameter of the primary silicon and the fraction of the fine area in dependence on the scan velocity for AlSi50 p. and AlSi50 Al + Si samples.

With an average diameter <5 µm in the coarse area and <2.5 µm in the fine area the additively manufactured primary Si was significantly smaller, compared to the platelike primary Si in the AlSi50 cast structure (>100 µm [4]), attributable to the high cooling rates in the PBF–LB/M process. In comparison to spray deposited AlSi50 (~13 µm [24]), the size of the primary Si particles could be further reduced. Furthermore, the primary Si sizes in the AlSi50 samples were in the range of the values recorded in the literature for PBF–LB/M, which were between 3 µm [10] and 7 µm [15]. Beyond that, in both areas, the primary Si size decreased with increasing scan velocity, which was probably due to the higher solidification rates at higher scan velocities [22]. In our case, the Si primary particle size showed no strong dependency on the powder variant, which contrasted with the findings by Garrard et al. [12], who reported about 18% larger primary Si particles in AlSi40 samples made of blended powders, compared to ones produced with pre-alloyed powder. The different process parameters used, and/or scanning strategy and/or the powder properties, could provide possible explanations for the discrepancy.

The fraction of the fine-grained areas, also shown in Figure 5, was 19–26% for the AlSi50 Al + Si and, hence, about 3% lower than for the AlSi50 p. + Si samples (22–30%). Again, this might be correlated to the laser absorption behavior of the different powder beds. The lower energy absorption by the Al + Si powder blend led to a reduced cooling rate inside the melting tracks and, finally, to a slightly coarser grain structure. In both cases, the fraction rose with increased scan velocity, caused by the higher solidification rates.

Summing up, the primary Si became smaller in all areas, as well as the fraction of the fine areas increasing at higher scan velocities, which correlated to an increased solidification rate. The results of both powder variants were very similar, and only the Al + Si blend resulted in a little lower fraction of fine areas.

3.3. *Mechanical Properties*

3.3.1. Hardness

The hardness values obtained from all AlSi50 and AlSi70 samples are shown in Figure 6. Firstly, it is obvious, that the hardness generally increased with increasing scan velocity, which was linked to the continuous refinement of the primary Si phase in the microstructure. The two AlSi50 variants exhibited comparable hardness values of ~190 HV5 at low scan velocities. This was in good agreement with the observations by Kang et al. [10], who measured a hardness of 188 HV0.3 in an AlSi50 sample produced with 500 mm/s.

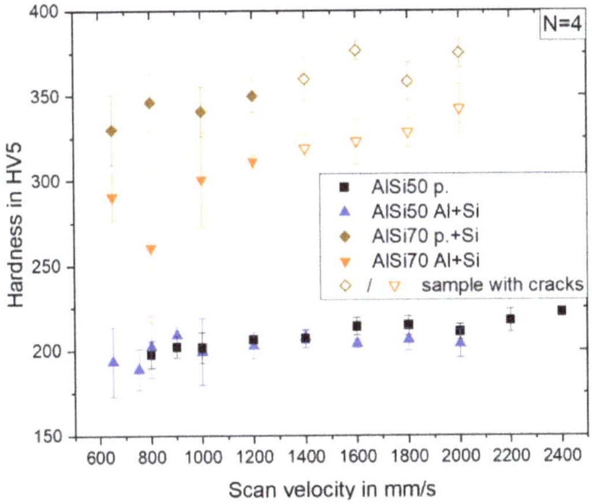

Figure 6. Vickers hardness vs. scan velocity for all AlSi50 and AlSi70 material variants.

Even with increasing scan velocity, both variants showed comparable hardness trends ending up with 222 HV5 for AlSi50 p. (at 2400 mm/s) and 204 HV5 for AlSi50 Al + Si (at 2000 m/s). Due to the higher Si content and the denser network of the Si primary phases, the hardness, considering only crack-free samples, and, thereby, a highest scanning velocity of 1200 mm/s, increased to 350 HV5 for AlSi70 p. + Si and 311 HV5 for AlSi70 Al + Si. Here, the samples built from the powder blend containing the pre-alloyed powder showed a significantly higher hardness in comparison to the one produced by the powder blend of Al + Si. This could be attributed to the higher contiguity of the primary Si phases, as shown in Figure 4d. A comparison to a published hardness value of a spray deposited AlSi70 alloy, which was significantly lower at 270 HV5 [7], underlines the attractiveness of the additive manufacturing approach for the production of these alloys.

3.3.2. Compressive Strength and Ductility

The course of the compressive strength–compression curves in Figure 7 can be divided into three parts. Firstly, the material deforms were merely elastic, so the compressive stress rose in a straight line. When reaching 0.2% compressive offset yield strength (0.2% OYS), the material started to deform plastically, e.g., at 450 MPa for the presented AlSi50 p. 1000 mm/s graph. In this section, the curve flattened, and the round bar started to bulge with advancing compression; but the compressive stress was still rising. In the last section, the compressive stress decreased, and the sample failed and sheared at a 45° angle to the compression direction. At this point, the compressive strength (CS) and fracture compression were reached, which were 590 MPa and 5.2% for the AlSi50 p. 1000 mm/s curve shown.

Analogous to the hardness, the 0.2% OYS and the CS also rose with increasing scan velocity, independent of the Si content and powder variant. Simultaneously, the fracture compression decreased. In the case of AlSi50 p., a rise in scan velocity from 1000 mm/s to 2400 mm/s caused an increment of the 0.2% OYS from 459 $\pm$ 27 MPa to 576 $\pm$ 35 MPa and a rise of CS from 584 $\pm$ 15 MPa to 742 $\pm$ 11 MPa, while the fracture compression declined from 5.6 $\pm$ 0.5% to 3.7 $\pm$ 0.3%.

With increasing scan velocity, finer grain structures formed, which increased the strength according to the Hall–Patch mechanism [25,26]. Moreover, the above-mentioned trends of the decreasing primary Si size and the higher fraction of fine area with increasing scan velocity could enhance this effect. Conditioned by the higher fraction of the fine area a decrease of the extensive matrix zones in the coarse areas occurred. As the matrix was more ductile than the primary Si, the reduced matrix areas resulted in lower ductility and fracture compression in the materials built with higher scan velocities. Furthermore, the matrix–particle interface enlarged due to the refinement of the primary Si particles. This hindered the dislocation movement, which also led to reduced ductility.

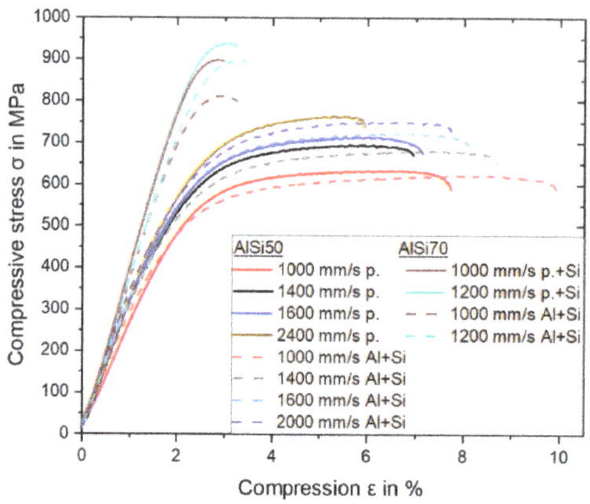

Figure 7. Compressive stress-compression diagram for samples of both AlSi50 variants fabricated with four scan velocities and for samples of both AlS70 variants produced with two scan velocities.

The obtained values were in good agreement with the findings by Jia et al. [4], who measured a CS of ~667 MPa and a fracture compression of ~4% in a sample built with a scan velocity of 1455 mm/s. The value for the 0.2% OYS of approximately 600 MPa was notably higher than the measured value of 475 ± 30 MPa at 1400 mm/s in this work. Since all process parameters used were quite comparable, one reason for the discrepancy could be the sample geometry. Jia et al. [4] used round bars with dimensions of Ø 6 mm × 25 mm, which corresponded to a height–diameter ratio >4 (in contrast to <2 used in this work). At this dimension, a buckling or bending of the samples cannot be ruled out, according to the DIN 50106:2016-11 [27], which could lead to a more inaccurate measurement.

Compared to the AlSi50 p. samples, the AlSi50 Al + Si samples showed lower 0.2% OYS, but similar CS. On the one hand, the lower contiguity of the primary Si led to a more ineffective load redistribution by the particles, which could cause higher stresses in the matrix of the AlSi50 Al + Si samples at the same external forces and activate dislocation sources [28]. This could be one reason for the inferior 0.2% OYS of the AlSi50 Al + Si samples. On the other hand, the larger matrix–particle interface could create a bigger backlog effect for the dislocations, leading to almost identical compressive strengths [29,30]. Furthermore, the AlSi50 Al + Si samples exhibited about 2% higher fracture compressions than the AlSi50 p. samples with a maximum of 7.3 ± 0.8% at 1000 mm/s. This could possibly be attributed to differences in the microstructure, like the particle morphology and the contiguity in the transition areas. In the AlSi50 p. samples, the primary Si particles were more often edgy, which could induce stress peaks decreasing ductility [12]. Additionally, the matrix size between the fine and coarse areas, which was larger in the transition area of the AlSi50 Al + Si samples, due to the less connected primary Si phases, could partly compensate the stress by plastic deformation.

In the same way as for the hardness, the CS also rose with increasing Si content. At the same scan velocity of 1000 mm/s the CS rose from 584 ± 15 MPa for AlSi50 p. and 595 ± 17 MPa for AlSi50 Al + Si to 897 ± 6 MPa for AlSi70 p. + Si and 805 ± 19 MPa for AlSi70 Al + Si. The strength increment was accompanied by a decrement in ductility with the result that the fracture compression reduced to 0.7 ± 0.1% for AlSi70 p. + Si and 1.2 ± 0.1% for AlSi70 Al + Si. The increment of the scan velocity up to 1200 mm/s, which was the highest velocity resulting in crack-free samples, resulted in no difference in the fracture compression, probably based on the small enhancement, or on the already low level of ductility. Apart from that, the scan velocity showed an influence on the strength, such that the CS increased to 935 ± 5 MPa for AlSi70 p. + Si and 870 ± 27 MPa for AlSi70 Al + Si with an increment from 1000 mm/s to 1200 mm/s.

The significant enhancement in strength from the AlSi50 to the AlSi70 samples could be attributed to the higher contiguity of the primary Si, which enabled more effective load redistribution on this skeletal structure and a shielding of the matrix [29,31]. As the contiguity was higher in the AlSi70 p. + Si samples, they showed superior strengths compared to the AlSi70 Al + Si samples. Otherwise, the dominant skeletal structure was extremely brittle resulting in the decrement of ductility compared to AlSi50, in which the primary Si phase was embedded in a more ductile Al and Si matrix.

To sum up the trends during compressive testing, strength was enhanced with increasing scan velocity and Si content but with a simultaneous reduction in ductility. In addition, the use of the blended powder could improve the ductility of Al-high Si alloys without (AlSi50), or with only small (AlSi70) loss of strength.

3.3.3. Young's Modulus

Due to the higher Young's modulus of Si, of approximately 160 GPa, compared to Al at about 70 GPa, the Young's modulus in Al-high Si alloys could be increased by increasing the Si content. This is illustrated in Figure 8 and by literature data on AlSi alloys, up to 40 wt.% Si. In addition, the measurement results of one AlSi50 p. sample and one AlSi70 Al + Si sample (both built with 1000 mm/s) of this work are shown. The Young's modulus increased from 93 GPa for AlSi40 [12] to 105 GPa for the AlSi50 p. sample, and further increased up to 124 GPa in the case of the AlSi70 Al + Si sample, which continued the trend evident in the literature data. Despite the fact that the trend did not proceed in an exact linear correlation, according to the rule of mixture, but in a curve below the straight line, it was possible to adjust, and also to further increase, the Young's modulus via accommodation of the Si content.

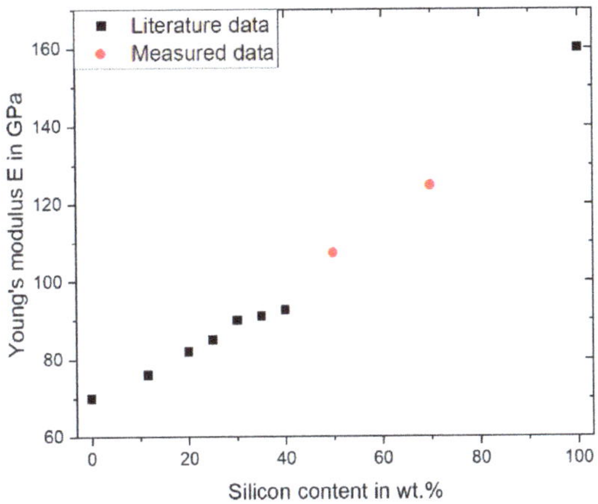

Figure 8. Literature data and measured data (AlSi50 p. and AlSi70 Al + Si from this work) for the Young's modulus depending on the silicon content. The literature data was for pure aluminum and silicon [24], AlSi10Mg [11], AlSi20, AlSi25, AlSi30, AlSi35 all from [32] and AlSi40 [12].

4. Conclusions

In this work, a fundamental study on the additive manufacturing of Al-high Si alloys (AlSi50 and AlSi70) by means of PBF–LB/M technology was carried out. It was demonstrated, for the first time, that it is possible to produce crack-free and highly dense samples, even up to an extremely high Si content of 70 wt.%, which was realized by adding elemental Si to pre-alloyed AlSi50 powder, or by directly mixing Al and Si.

The porosity of the samples strongly depended on the scan velocity and was influenced only in a minor way by the powder source used. However, ageing of the powders could lead to massive insertion of small pores of <20 μm in diameter, which were only visible by microscopic analysis. This should be considered during alloy development with respect to pore affected properties.

The mechanical properties of the Al-high Si alloys were clearly correlated to the microstructure. Higher scan velocities led to a decrease in grain size and, consequently, to an increase in hardness, compression strength and Young's Modulus. A significant improvement could be achieved by increasing the Si content from 50 to 70 wt.%, inducing the formation of a dense network of primary Si phases. On the other hand, some loss of ductility existed, which could be partly prohibited using elemental powder blends.

Presumably the cracking issue, which sporadically appeared in the AlSi70 samples at higher scan velocities, is the main issue for establishing AM of Al-high Si alloys with extremely high Si content. Hence, further development of adequate countermeasures, especially in the course of upscaling the sample size, are required.

Author Contributions: Conceptualization, J.H.R. and M.T.; Investigation, J.H.R., F.H. and D.B.; Methodology, J.H.R., F.H. and D.B.; Project administration, M.T. and C.R.; Resources, H.W.H. and M.S.; Supervision, H.W.H., M.S. and J.F.; Validation, J.H.R. and M.T.; Writing—original draft, J.H.R. and M.T.; Writing—review & editing, F.H., C.R. and J.F. All authors have read and agreed to the published version of the manuscript.

Funding: This research received no external funding.

Institutional Review Board Statement: Not applicable.

Informed Consent Statement: Not applicable.

Data Availability Statement: Not applicable.

Conflicts of Interest: The authors declare no conflict of interest.

References

1. Heidler, N.; Von Lukowicz, H.; Hilpert, E.; Risse, S.; Alber, L.; Klement, J.; Heine, F.; Bölter, R.; Armengol, J.M.P. Topology optimization and additive manufacturing of an optical housing for space applications. *EPJ Web Conf.* **2019**, *215*, 01005. [CrossRef]
2. Jacobson, D.M.; Ogilvy, A.J.W. Spray-deposited Al-Si (Osprey CE) alloys and their properties. *Mater. Werkst.* **2003**, *34*, 381–384. [CrossRef]
3. Liang, S.-M.; Schmid-Fetzer, R. Phosphorus in Al–Si cast alloys: Thermodynamic prediction of the AlP and eutectic (Si) solidification sequence validated by microstructure and nucleation undercooling data. *Acta Mater.* **2014**, *72*, 41–56. [CrossRef]
4. Jia, Y.D.; Zhang, L.B.; Ma, P.; Scudino, S.; Wang, G.; Yi, J.; Eckert, J.; Prashanth, K.G. Thermal expansion behavior of Al–xSi alloys fabricated using selective laser melting. *Prog. Addit. Manuf.* **2020**, *5*, 247–257. [CrossRef]
5. Yu, K.; Li, C.; Yang, J.; Cai, Z.Y. Production and properties of a 90% Si-Al alloy for electronic packaging applications. *MSF* **2009**, *610–613*, 542–545. [CrossRef]
6. Cai, Z.; Zhang, C.; Wang, R.; Peng, C.; Wu, X.; Li, H. Microstructure, mechanical and thermo-physical properties of Al–50Si–xMg alloys. *Mater. Sci. Eng. A* **2018**, *730*, 57–65. [CrossRef]
7. Wang, F.; Xiong, B.; Zhang, Y.; Zhu, B.; Liu, H.; Wei, Y. Microstructure, thermo-physical and mechanical properties of spray-deposited Si–30Al alloy for electronic packaging application. *Mater. Charact.* **2008**, *59*, 1455–1457. [CrossRef]
8. Aboulkhair, N.T.; Simonelli, M.; Parry, L.; Ashcroft, I.; Tuck, C.; Hague, R. 3D printing of Aluminium alloys: Additive manufacturing of aluminium alloys using selective laser melting. *Prog. Mater. Sci.* **2019**, *106*, 100578. [CrossRef]
9. Wang, Z.; Ummethala, R.; Singh, N.; Tang, S.; Suryanarayana, C.; Eckert, J.; Prashanth, K.G. Selective laser melting of aluminum and its alloys. *Materials* **2020**, *13*, 4564. [CrossRef]
10. Kang, N.; Coddet, P.; Chen, C.; Wang, Y.; Liao, H.; Coddet, C. Microstructure and wear behavior of in-situ hypereutectic Al–high Si alloys produced by selective laser melting. *Mater. Des.* **2016**, *99*, 120–126. [CrossRef]
11. Aboulkhair, N.T.; Maskery, I.; Tuck, C.; Ashcroft, I.; Everitt, N.M. The microstructure and mechanical properties of selectively laser melted AlSi10Mg: The effect of a conventional T6-like heat treatment. *Mater. Sci. Eng. A* **2016**, *667*, 139–146. [CrossRef]
12. Mueller, M.; Riede, M.; Eberle, S.; Reutlinger, A.; Brandão, A.D.; Pambaguian, L.; Seidel, A.; López, E.; Brueckner, F.; Beyer, E.; et al. Microstructural, mechanical, and thermo-physical characterization of hypereutectic AlSi40 fabricated by selective laser melting. *J. Laser Appl.* **2019**, *31*, 22321. [CrossRef]

13. Garrard, R.; Lynch, D.; Carter, L.N.; Adkins, N.J.; Gie, R.; Chouteau, E.; Pambaguian, L.; Attallah, M.M. Comparison of LPBF processing of AlSi40 alloy using blended and pre-alloyed powder. *Addit. Manuf. Lett.* **2022**, *2*, 100038. [CrossRef]
14. Eberle, S.; Reutlinger, A.; Curzadd, B.; Mueller, M.; Riede, M.; Wilsnack, C.; Brandão, A.; Pambaguian, L.; Seidel, A.; López, E.; et al. Additive manufacturing of an AlSi40 mirror coated with electroless nickel for cryogenic space applications. In Proceedings of the International Conference on Space Optics—ICSO 2018, Chania, Greece, 9–12 October 2018; p. 40, ISBN 9781510630772.
15. Hanemann, T.; Carter, L.N.; Habschied, M.; Adkins, N.J.; Attallah, M.M.; Heilmaier, M. In-situ alloying of AlSi10Mg+Si using Selective Laser Melting to control the coefficient of thermal expansion. *J. Alloys Compd.* **2019**, *795*, 8–18. [CrossRef]
16. Jia, Y.D.; Ma, P.; Prashanth, K.G.; Wang, G.; Yi, J.; Scudino, S.; Cao, F.Y.; Sun, J.F.; Eckert, J. Microstructure and thermal expansion behavior of Al-50Si synthesized by selective laser melting. *J. Alloys Compd.* **2017**, *699*, 548–553. [CrossRef]
17. Weingarten, C.; Buchbinder, D.; Pirch, N.; Meiners, W.; Wissenbach, K.; Poprawe, R. Formation and reduction of hydrogen porosity during selective laser melting of AlSi10Mg. *J. Mater. Process. Technol.* **2015**, *221*, 112–120. [CrossRef]
18. De Terris, T.; Andreau, O.; Peyre, P.; Adamski, F.; Koutiri, I.; Gorny, C.; Dupuy, C. Optimization and comparison of porosity rate measurement methods of Selective Laser Melted metallic parts. *Addit. Manuf.* **2019**, *28*, 802–813. [CrossRef]
19. Zhang, B.; Li, Y.; Bai, Q. Defect formation mechanisms in selective laser melting: A review. *Chin. J. Mech. Eng.* **2017**, *30*, 515–527. [CrossRef]
20. Gong, H.; Nadimpalli, V.K.; Rafi, K.; Starr, T.; Stucker, B. Micro-CT evaluation of defects in Ti-6Al-4V parts fabricated by metal additive manufacturing. *Technologies* **2019**, *7*, 44. [CrossRef]
21. Kempen, K.; Vrancken, B.; Thijs, L.; Buls, S.; Van Humbeeck, J.; Kruth, J.P. Lowering thermal gradients in selective laser melting by pre-heating the baseplate. In Proceedings of the Solid Freeform Fabrication Symposium, Austin, TX, USA, 12–15 August 2013.
22. Qin, H.; Fallah, V.; Dong, Q.; Brochu, M.; Daymond, M.R.; Gallerneault, M. Solidification pattern, microstructure and texture development in laser powder bed fusion (LPBF) of Al10SiMg alloy. *Mater. Charact.* **2018**, *145*, 29–38. [CrossRef]
23. Khorasani, M.; Ghasemi, A.; Leary, M.; Sharabian, E.; Cordova, L.; Gibson, I.; Downing, D.; Bateman, S.; Brandt, M.; Rolfe, B. The effect of absorption ratio on meltpool features in laser-based powder bed fusion of IN718. *Opt. Laser Technol.* **2022**, *153*, 108263. [CrossRef]
24. Jia, Y.; Cao, F.; Scudino, S.; Ma, P.; Li, H.; Yu, L.; Eckert, J.; Sun, J. Microstructure and thermal expansion behavior of spray-deposited Al–50Si. *Mater. Des.* **2014**, *57*, 585–591. [CrossRef]
25. Amirjan, M.; Sakiani, H. Effect of scanning strategy and speed on the microstructure and mechanical properties of selective laser melted IN718 nickel-based superalloy. *Int. J. Adv. Manuf. Technol.* **2019**, *103*, 1769–1780. [CrossRef]
26. Esmaeilizadeh, R.; Keshavarzkermani, A.; Ali, U.; Mahmoodkhani, Y.; Behravesh, B.; Jahed, H.; Bonakdar, A.; Toyserkani, E. Customizing mechanical properties of additively manufactured Hastelloy X parts by adjusting laser scanning speed. *J. Alloys Compd.* **2020**, *812*, 152097. [CrossRef]
27. DIN. *Deutsches institur für normung e.v. Prüfung Metallischer Werkstoffe—Druckversuch Bei Raumtemperatur*; Beuth Verlag GmbH: Berlin, Germany, 2016.
28. German, R.M. *Particulate Composites: Fundamentals and Applications*; Springer: Cham, Switzerland, 2016; ISBN 978-3-319-29917-4.
29. Huang, L.; Geng, L. *Discontinuously Reinforced Titanium Matrix Composites: Microstructure Design and Property Optimization*; Springer: Singapore, 2017; ISBN 978-981-10-4447-2.
30. Dieter, G.E. *Mechanical Metallurgy, SI Metric Ed.*; McGraw-Hill: London, UK, 2001; ISBN 9780071004060.
31. Kim, C.-S.; Massa, T.R.; Rohrer, G.S. Modeling the influence of orientation texture on the strength of Wc-co composites. *J. Am. Ceram. Soc.* **2007**, *90*, 199–204. [CrossRef]
32. Nikanorov, S.P.; Volkov, M.P.; Gurin, V.N.; Burenkov, Y.; Derkachenko, L.I.; Kardashev, B.K.; Regel, L.L.; Wilcox, W.R. Structural and mechanical properties of Al–Si alloys obtained by fast cooling of a levitated melt. *Mater. Sci. Eng. A* **2005**, *390*, 63–69. [CrossRef]

materials

MDPI

Article

Influence of the AlSi7Mg0.6 Aluminium Alloy Powder Reuse on the Quality and Mechanical Properties of LPBF Samples

Irina Smolina [1,2,*], Konrad Gruber [1], Andrzej Pawlak [1,2], Grzegorz Ziółkowski [1], Emilia Grochowska [1], Daniela Schob [2], Karol Kobiela [1], Robert Roszak [2,3], Matthias Ziegenhorn [2] and Tomasz Kurzynowski [1]

1 Centre for Advanced Manufacturing Technologies (CAMT-FPC), Faculty of Mechanical Engineering, Wroclaw University of Science and Technology, ul. Łukasiewicza 5, 50-371 Wroclaw, Poland; konrad.gruber@pwr.edu.pl (K.G.); andrzej.p.pawlak@pwr.edu.pl (A.P.); grzegorz.ziolkowski@pwr.edu.pl (G.Z.); emilia.grochowska@pwr.edu.pl (E.G.); karol.kobiela@pwr.edu.pl (K.K.); tomasz.kurzynowski@pwr.edu.pl (T.K.)

2 Chair of Engineering Mechanics and Machine Dynamics, Brandenburg University of Technology Cottbus-Senftenberg, Universitätsplatz 1, 01968 Senftenberg, Germany; daniela.schob@b-tu.de (D.S.); robert.roszak@b-tu.de (R.R.); matthias.ziegenhorn@b-tu.de (M.Z.)

3 Institute of Applied Mechanics, Poznan University of Technology, Pl. M. Sklodowskiej-Curie 5, 60-965 Poznan, Poland

* Correspondence: iryna.smolina@pwr.edu.pl; Tel.: +48-71-320-42-09

Citation: Smolina, I.; Gruber, K.; Pawlak, A.; Ziółkowski, G.; Grochowska, E.; Schob, D.; Kobiela, K.; Roszak, R.; Ziegenhorn, M.; Kurzynowski, T. Influence of the AlSi7Mg0.6 Aluminium Alloy Powder Reuse on the Quality and Mechanical Properties of LPBF Samples. *Materials* **2022**, *15*, 5019. https://doi.org/10.3390/ma15145019

Academic Editors: Mostafa Hassani, Hongze Wang and Greta Lindwall

Received: 9 June 2022
Accepted: 16 July 2022
Published: 19 July 2022

Abstract: Additive manufacturing (AM) is dynamically developing and finding applications in different industries. The quality of input material is a part of the process and of the final product quality. That is why understanding the influence of powder reuse on the properties of bulk specimens is crucial for ensuring the repeatable AM process chain. The presented study investigated the possibility of continuous reuse of AlSi7Mg0.6 powder in the laser powder bed fusion process (LPBF). To date, there is no study of AlSi7Mg0.6 powder reuse in the LPBF process to be found in the literature. This study aims to respond to this gap. The five batches of AlSi7Mg0.6 powder and five bulk LPBF samples series were characterised using different techniques. The following characteristics of powders were analysed: the powder size distribution (PSD), the morphology (scanning electron microscopy—SEM), the flowability (rotating drum analysis), and laser light absorption (spectrophotometry). Bulk samples were characterised for microstructure (SEM), chemical composition (X-ray fluorescence spectrometry—XRF), porosity (computed tomography—CT) and mechanical properties (tensile, hardness). The powder was reused in subsequent processes without adding (recycling/rejuvenation) virgin powder (collective ageing powder reuse strategy). All tested powders (powders P0–P4) and bulk samples (series S0–S3) show repeatable properties, with changes observed within error limits. Samples manufactured within the fifth reuse cycle (series S4) showed some mean value changes of measured characteristics indicating initial degradation. However, these changes also mostly fit within error limits. Therefore, the collective ageing powder reuse strategy is considered to give repeatable LPBF process results and is recommended for the AlSi7Mg0.6 alloy within at least five consecutive LPBF processes.

Keywords: additive manufacturing; powder reuse; aluminium alloy; porosity; mechanical properties

1. Introduction

Additive manufacturing (AM) is dynamically developing and finds applications in industries such as aerospace, automotive, and medicine. The main benefits of AM are the freedom of design, a low level of waste, a decrease in the number of technological operations, the production time, and the supply chain cost in low series production [1,2].

However, certain limits slow down the complete implementation of AM technologies in the industry. For example, a lack of standardization, problems with repeatability, the possibility of quickly moving the process from one machine to another, and insufficient knowledge about powder reuse.

Powder reuse is critically essential for laser powder bed fusion (LPBF). According to ISO/ASTM 52900:2021 [3], LPBF is used to produce objects from powdered materials and uses one or more lasers to fuse or melt the deposited layers of powder particles selectively. Therefore, the possibility of reusing powder in more than one process makes PBF technologies more sustainable and decreases production costs. It is an accepted practice to use the unmelted powder material more than once for the process. Usually, for powder reuse in LPBF, manufacturers use powder recycling/regeneration procedures. However, it is a time- or number-limited usage due to quality and material concerns. Besides that, the authors in the paper [4] explain the difference between recycling (rejuvenation) and reuse. They are limiting the second term to the repetitive use of the same powder, without or with minimal post-processing, such as sieving/screening.

Many authors [5–8] have been investigating the effect of powder particle size and oxygen content during the LPBF process on the results of its processing. All the researchers agreed that the powder used in consecutive processes must be free of contamination, e.g., oxidation and cross-contamination, which can occur accidentally by mixing metal powders of different chemical compositions [9]. Even spatters formed during the metal powder processing can be treated as contamination, even though the chemical composition remains the same. However, the oxygen content very often increases [10]. These findings push the researchers working on process LPBF development to consider the effect of powder reuse, applied strategies of operating powder between the processes, and the changes of powder characteristics on the result of the AM process.

Various strategies for reusing powders for LPBF can also be found in the literature. For example, the authors of [11] described two of them: (1) continuous refreshing and (2) collective ageing. The first approach states that the powder volume for each build job should be the same. Therefore, all remaining powder is filled with fresh powder after the process, eliminating the powder losses caused by manufactured parts and supports, powder loss due to spatter formation, and machine operation. On the other hand, the collective ageing strategy is about using the once-loaded powder in a machine till the remaining powder amount is insufficient for the next job. Then, the remaining powder volume is filled with a used powder that was sieved and loaded into the main machine tank. Both methods are different in terms of how much powder can be operated in total and how much powder is needed to sieve and mix after every process.

There are many studies on powder reuse in LPBF, including nickel-based alloys [10], iron-based alloys [12], titanium-based alloys [13], and aluminium-based alloys. Recently, published works on aluminium-based alloys focus on the AlSi10Mg [14] and Al-Si-Sc-Zr [15]. The main findings of [15] have pertained to the mechanical properties (ultimate tensile strength (UTS) and elongation at the break of specimens, respectively, built with virgin and reused powder are 565 MPa, 13% and 537 MPa, 11%) and porosity (0.06% for samples made from virgin powder compared to 0.15% for samples made for the reused powder). In addition, the authors of [16] analysed properties of AlSi10Mg alloy aged in air and used for direct-energy deposition (DED); they found out that the increase in oxygen content influences the processability of the powder and the properties of the final specimens.

Aluminium alloys cover about 1/4 of the AM powder market by volume [17]. Therefore, the interest in the effective use of these materials and the possibility of reusing the aluminium alloy powders remains at a high level [4]. Understanding the powder degradation mechanism of Al alloy in the LPBF process is essential. The AlSi7Mg0.6 alloy is gaining more and more attention in the AM industry [18–20]. However, to date, there is no study of AlSi7Mg0.6 powder reuse in the LPBF process to be found in the literature. This study aims to respond to this gap.

The goal of this study is to evaluate the influence of AlSi7Mg0.6 powder reuse on the material properties of LPBF samples. The powder will be reused in subsequent processes without adding (reuse/rejuvenation step) virgin/fresh powder, using only a double sieving procedure to eliminate the oversized powder particles.

The objectives of this study are:

- To verify the eventual changes in powder morphology and chemical composition during the continuous reuse;
- To evaluate the influence of eventual powder degradation on the quality of LPBF bulk specimens.

2. Materials and Methods

2.1. Materials and Processing

The AlSi7Mg0.6 powder with the chemical composition as listed in Table 1 is used. The powder is supplied by SLM Solutions Group AG (Lübeck, Germany).

Table 1. Chemical composition of the AlSi7Mg0.6 powder used in this study.

Standard	Al	Si	Mg	Ti	Fe	Cu	Mn	Zn	Other Total
EN AC42200 acc. to EN-1706, wt.%	Bal.	6.5–7.5	0.45–0.70	-	0.15–0.19	0.03–0.05	0.1	max 0.07	0.10
SLM Solutions, wt.%	Bal.	6.5–7.5	0.45–0.70	0.25	0.19	0.05	-	max 0.07	0.10

Five successive LPBF process cycles are carried out using the SLM 280 2.0 machine (SLM Solutions Group AG, Lübeck, Germany). The LPBF system used in this research is equipped with a 1070 nm fibre laser with a max. power of 700 W and a beam focus diameter between 80 and 115 μm. Each process is held under a protective atmosphere of pure argon (O_2 level kept below 100 ppm, argon purity class 5.0). In each process, six plates (150 mm × 30 mm × 4 mm) and twelve cylinders (Ø12 mm × 150 mm) are manufactured, as shown in Figure 1a,b. A build volume reduction (100 mm × 100 mm × 150 mm) is used to allow high-volume powder consumption. Parts are manufactured on a 1xxx series aluminium build platform with 98 mm × 98 mm × 20 mm dimensions, heated up to 150 °C and kept at this temperature during the LPBF process. Cylindrical samples are used for microstructure studies, and bar samples are used for tensile testing. To manufacture the samples, previously tested process parameters are used, in line with machine and powder supplier recommendations (SLM Solutions Group AG, Lübeck, Germany). As stated by the supplier, the process parameters should provide sample densities above 99% [21]. The same set of parameters is used for each of the five LPBF processes.

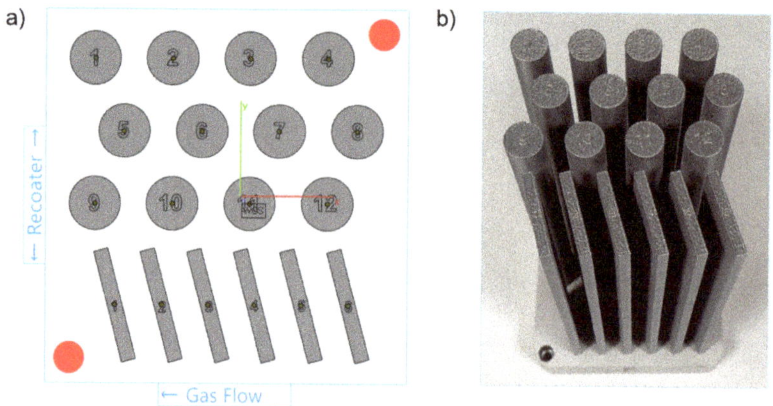

Figure 1. (**a**) LPBF process build job layout; (**b**) An exemplar build plate with samples manufactured.

The powder as received from the supplier after initial sieving through 75 μm sieve size is labelled as 'virgin' (powder P0), and the powder after each following process cycle is labelled as 'powder PN' (powders P1, P2, P3, and P4), where N is a number of powder reuse cycles prior the N process (Table 2). Therefore, the LPBF samples are appropriately named S0, S1 ... S4 (in line with the used powder batches P0, P1 ... P4). In addition, an oversized powder collected from the sieve was also characterized, labelled as PW (waste powder).

Table 2. The list of powder samples used in the research.

Powder Sample	Description
P0	Initial batch of virgin powder. P0 powder is dried and sieved before use.
P1, P2, P3, P4	Powder after 1, 2, 3, and 4 LPBF process cycles and double sieving.
PW	Waste powder. The powder that stayed on the sieves after double sieving.

The powder volume is not refilled throughout the experiments. This approach is named "continuous reuse" according to ASTM F3456-22 [22]. Initially, 15 kg of virgin AlSi7Mg0.6 powder was used to manufacture the first batch of samples (S0). After each process, the whole powder volume is removed from the machine and double sieved to eliminate the oversized powder particles. The whole powder volume collected after the N process is sieved in the first step, and then the second sieving is performed only for the overflown powder separated in the first sieving. Finally, the powder that passed through the sieves after the first and second sieving operations is mixed. A powder sieving station PSM 100 (SLM Solutions Group AG, Germany), equipped with a 75 μm flow sieve, is used to sieve the powder.

The total volume of processed powder decreases every build job due to the use for sample manufacturing and rejection after double sieving. The change in powder weight throughout the process cycles is shown in Figure 2. Finally, the height of the last build job has decreased to 120 mm. At the end of the experiment, the weight of powder remaining in use is 11 kg. The part weight to the powder in use weight ratio is about 1:10 and depends on the 'N' processing cycle (process 0 to process 4).

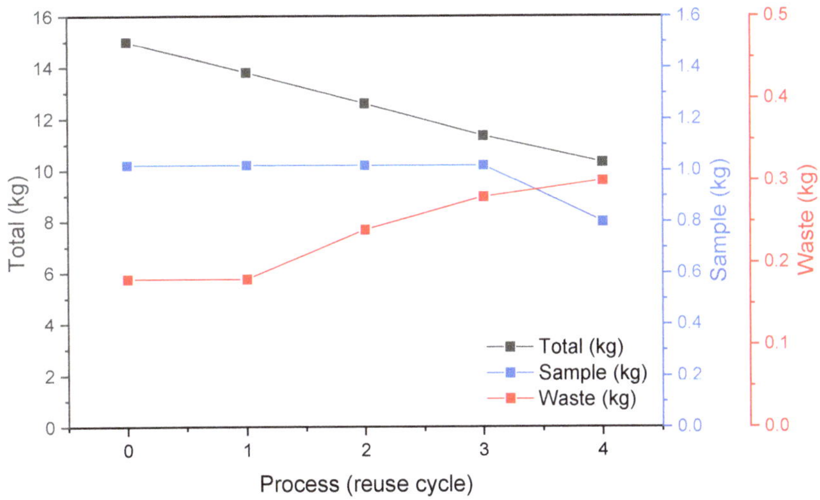

Figure 2. Summary of the total powder weight used in the study, including the weight of obtained samples and the weight of waste powder after double sieving.

Dogbone samples according to ASTM E8/E8M-16 [23] for mechanical testing are machined from the plates produced in each LPBF process (Figure 3). The length of tensile test samples is 100 mm, so the decreased build job height of the last LPBF process does not influence the size of tensile samples.

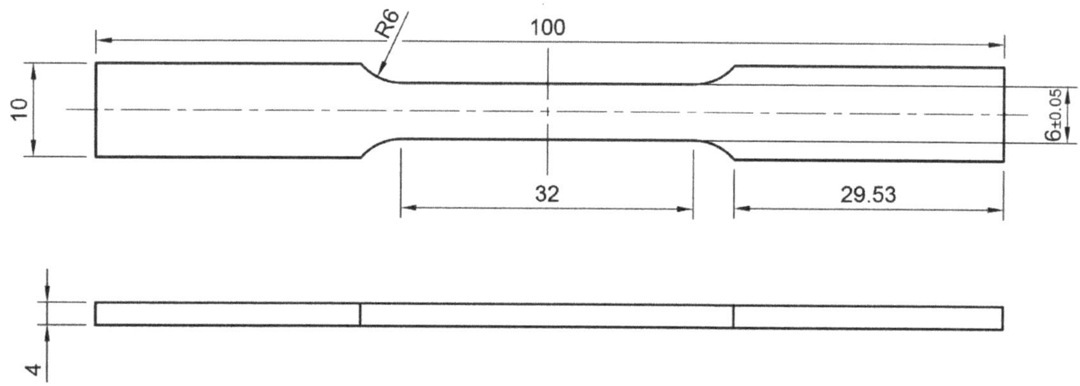

Figure 3. Tensile specimens' geometry according to ASTM E8/E8M-16 (all dimensions are in mm).

2.2. *Powder Characterization*

After each process cycle (LPBF + double sieving), the powder is characterised by morphology, flowability, and physicochemical properties. Finally, the results are compared with the properties of waste powder.

2.2.1. Powder Morphology

To measure the particle size distribution (PSD), the HELOS BR R4 + RODOS laser diffraction system, equipped with a VIBRI dispersion unit, is used (Sympatec GmbH, Clausthal-Zellerfeld, Germany). A 68-mbar vacuum and a 2-bar feed pressure are used to disperse particles during testing. A 70% feed rate and 1.5 mm gap width parameters of the VIBRI unit are used to feed the powder. Statistical analysis of the PSD is performed in the PAQXOS 3.1 software (Sympatec GmbH, Clausthal-Zellerfeld, Germany) according to the ISO 13320-1/ASTM B822-17 standards.

The individual powder particles' morphology and surface condition are characterised by microscopic investigation using SEM EVO MA25 scanning electron microscope (CARL ZEISS, Oberkochen, Germany). The procedure is performed according to ISO 13322-1 standard.

2.2.2. Powder Flowability

A rotating drum (GranuDrum) granular flow analyser is used (Granutools, Awans, Belgium) to characterise powder flowability. The instrument is an automated tester providing the cohesion analysis within the powder flowing in a rotating drum. First, a flowing powder interface position snapshots are analysed. Based on this, the cohesive index is derived. The higher the powder fluctuation during rotation flow, the higher the cohesive index. In addition, the mean avalanche angle is measured during the test. Cohesive index and mean avalanche angles are determined for the drum's increasing and decreasing rotational speed (hysteresis mode), i.e., 1, 2, 5, 10, 20, 30, 40, 50, and 60 RPM. Thirty flowing powder images are taken at each RPM level with a 1 Hz sampling rate to calculate the cohesive index and mean avalanche angle.

2.2.3. Laser Absorption

Laser absorption assessment is performed with the spectrophotometry method using Exemplar Plus BTC655N-ST laser radiation absorption spectrophotometer (B&W Tek, Newark, DE, USA). The powder sample is placed inside the integrating sphere. Modulated monochromatic light (in a range of 900–1100 nm) is shined at the powder sample. Light reflection is compared to a reference sample with almost 100% reflectivity. Based on the measurement, laser absorption is calculated.

The changes in absorption can be directly translated into the course of the LPBF process and the need to adjust the process parameters (process window) to the current state of the powder.

2.2.4. Chemical Composition

Chemical composition is evaluated using an energy dispersive X-ray fluorescence spectrometer (ED-XRF) SPECTRO XEPOS (METEK, Kleve, Germany). The measurement is done for powder and bulk specimens. The measurement was repeated 3x for one random bulk specimen from each series (S0 ... S4) and for each powder type (P0 ... P4). Bulk specimens from the XZ plane are prepared as a metallographic sample (polished and ground).

2.3. Sample Characterization

2.3.1. Porosity

The quality of the as-built samples is evaluated using a technical computed tomography method. X-ray computed tomography (XCT) enables the reconstruction and evaluation of the external and internal structure of the manufactured samples, which is especially important for samples produced with additive technologies [24,25]. The volumetric models obtained as a result of the XCT reconstruction allow to determine the volume of voids or pores filled with incompletely melted powder (V_{por}) and the volume of the melted powder (V_m), making it possible to determine the volumetric porosity (P) according to the Equation (1) [26]:

$$P\,[\%] = \frac{V_{por}}{V_m + V_{por}} \times 100\% \qquad (1)$$

The XCT system phoenix v | tome | x m 300/180 (GE Sensing & Inspection Technologies GmbH, Wunstorf, Germany) is used in the study. A micro-focus X-ray tube with a parameter setting (voltage 160 kV and current 120 μA) is used to X-ray the samples. Such parameters of the X-ray tube with a 2K flat-panel digital detector (10-bit) allowed for scanning a set of ROIs (region of interest covering the gauge section) for six samples with a resolution (voxel size) of 19.79 μm. The reconstruction is carried out using dedicated software phoenix datos | x 2.7.2 (GE Sensing & Inspection Technologies GmbH, Wunstorf, Germany) with measurement artefact correction (ring artefact, axis alignment, beam hardening) and noise reduction. Data processing, including data thresholding and porosity detection, is performed using software VG Studio MAX 3.3 (Volume Graphics GmbH, Heidelberg, Germany).

2.3.2. Microstructure Characterization

Metallographic specimens are prepared to analyse the microstructure. The microstructure analysis is carried out on a cross-section parallel to the build direction. The surface of the metallographic specimens is etched with Kroll's reagent with the following composition:

$$68\ cm^3\ H_2O + 16\ cm^3\ HNO_3 + 16\ cm^3\ HF.$$

The confocal laser scanning microscope (CLSM) LEXT OLS4000 (Olympus, Tokyo, Japan) and the Zeiss SEM EVO MA25 (CARL ZEISS, Oberkochen, Germany) scanning electron microscope (SEM), equipped with an EDS (energy dispersive spectroscopy) analysis system, are used to capture microstructure images.

2.3.3. Mechanical Properties

The static tensile samples are designed following ASTM E8/E8M-16a (room temperatures) [23]. Tensile tests are performed on the HC-25 ZwickRoell servohydraulic testing machine (ZwickRoell GmbH & Co. KG, Ulm, Germany) with a test frame using a ±25 kN load-cell. The tests are carried out with a strain rate of 0.0008 s^{-1} and are continued until the sample brake. Five specimens are used for each series (S0, S1, S2, S3, and S4).

Hardness tests using a Zwick Roel ZHVμ-A hardness tester (Zwick-Roell, Leominster, United Kingdom) are performed. Vickers hardness cross-section profiles are determined at a 2.94 N load (300 g). Five indents are made on each sample used for microstructure evaluation, and a mean value is calculated for each tested sample.

3. Results

3.1. Powder Characterization

3.1.1. Powder Morphology

In the virgin state (P0), aluminium powder (AlSi7Mg0.6) is characterized by a "potato"-like shape and has some satellites (Figure 4a). Powder samples after each run of the LPBF process (P1–P4, Figure 4b–e) are characterized by a similar morphology and demonstrate a lack of changes due to processing. However, the waste powder (PW) is significantly different—it has a different surface morphology, particles are 2–3× bigger than virgin powder particles, and lacks satellites.

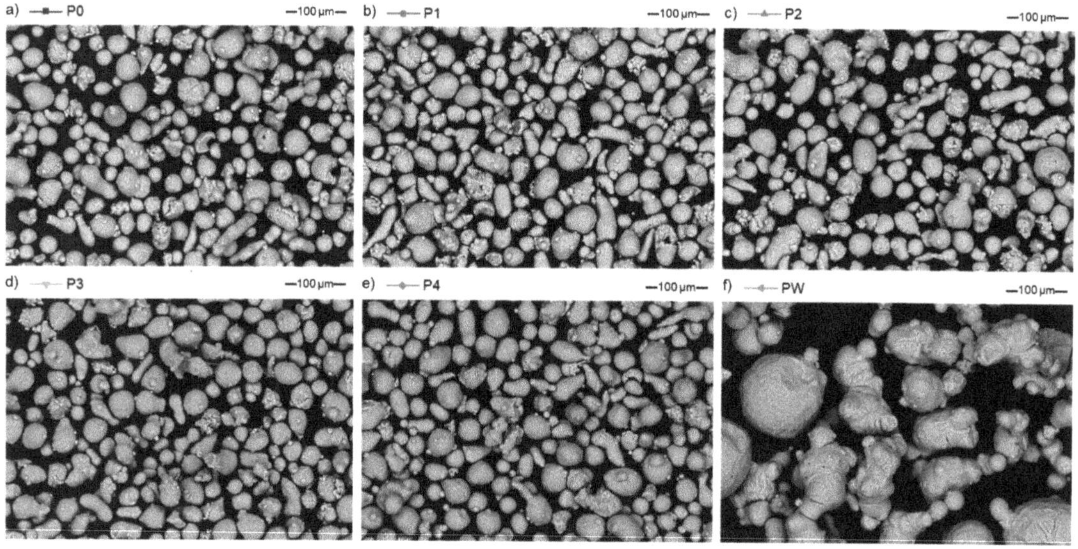

Figure 4. Microscopic images of the AlSi7Mg0.6 powders in various processing states, SEM/BSE: (**a**) virgin (P0); (**b**) after 1st LPBF process (P1); (**c**) after 2nd LPBF processes (P2); (**d**) after 3rd LPBF processes (P3); (**e**) after 4th LPBF processes (P4); and (**f**) waste (PW).

The obtained particle size distributions (PSD) for each evaluated powder are comparable. There is no significant difference between each consecutive sieving cycle (P0 to P4). The curves for powders P0–P4 are plotted simultaneously one on another, both for distribution density q3 (Figure 5a) and the cumulative density Q3 (Figure 5b), showing almost identical plots. The SEM observations and PSD analysis show the presence of satellite particles throughout the powder states with no variations. The values of $x_{10,3}$, $x_{50,3}$, and $x_{90,3}$ characteristic particles size parameters are presented in Table 3. In the case of the PW probe, PSD curves are clearly visible and are moved to the right side, which means that the powder sample consists of much larger diameter particles. PW $x_{50,3}$ is 2.5-times higher than P0–P4 and $x_{90,3}$ is 3.5-times higher than P0–P4.

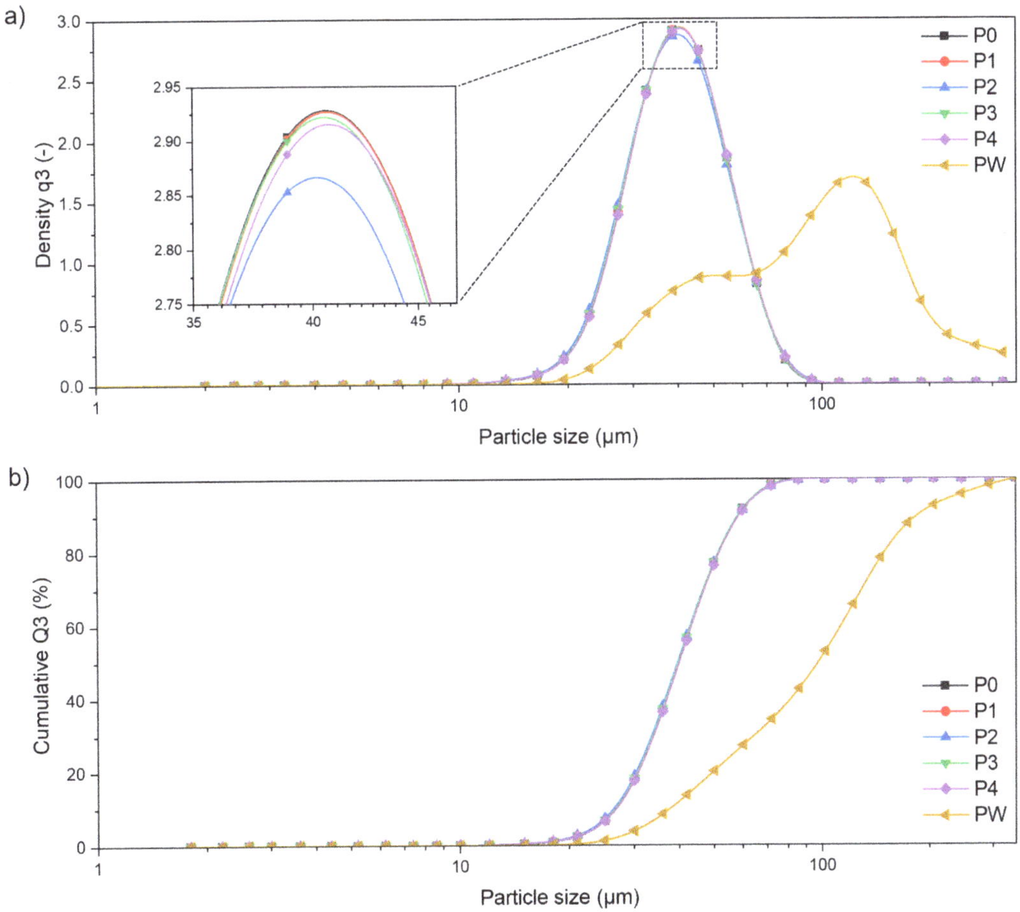

Figure 5. Particle size distribution of AlSi7Mg0.6 for each of the powder states: (**a**) distribution density q3; (**b**) cumulative density Q3.

Table 3. Particle size distribution of AlSi7Mg0.6—volume-weighted characteristic values obtained in the laser diffraction measurements for each of the powder states.

Parameter	P0 (μm)	P1 (μm)	P2 (μm)	P3 (μm)	P4 (μm)	PW (μm)
$x_{10,3}$	26.50	26.50	26.14	26.40	26.54	37.78
$x_{50,3}$	40.02	40.07	39.76	39.94	40.21	97.50
$x_{90,3}$	58.75	58.85	58.97	58.73	59.08	188.34

The PS distribution of PW is not a normal distribution, unlike the PSD of P0–P4. This suggests that the waste powder is not homogeneous in its volume. Based on this observation, the waste powder (PW) is a mixture of a non-remelted powder, and a re-melted spatter powder.

3.1.2. Powder Flow Properties

The measured flow properties of the powders in each of the P(N) states behave in a very similar way. The measured flow properties of the powders in each of the P(N) states behave in a very similar way. All powder samples are characterized by a relatively high cohesive index and a tendency to increase a cohesive index with an increasing drum speed. According to [27], it is influenced by particle size and shape. An increasing drum speed increases bonding between the particles and influences static (angle of repose) and dynamic (cohesive index and flow) behaviour.

Based on the plots in Figure 6, there is no difference between P0 and P4 powders. Therefore, only the curve for the PW sample can be differentiated from the reused powders. The cohesive index for the waste powder slightly increases up to 10 RPM, and with an increase in the rotational speed, the cohesive index remains constant (Figure 6a). When analysing the trend of the avalanche angle for the PW sample, again, 10 RPM is a threshold where the behaviour changes, and below 10 RPM, the avalanche angle is constant. With a higher rotation speed, the measured avalanche angle is comparable to the sieved P0–P4 powders used in the consecutive processes (Figure 6b).

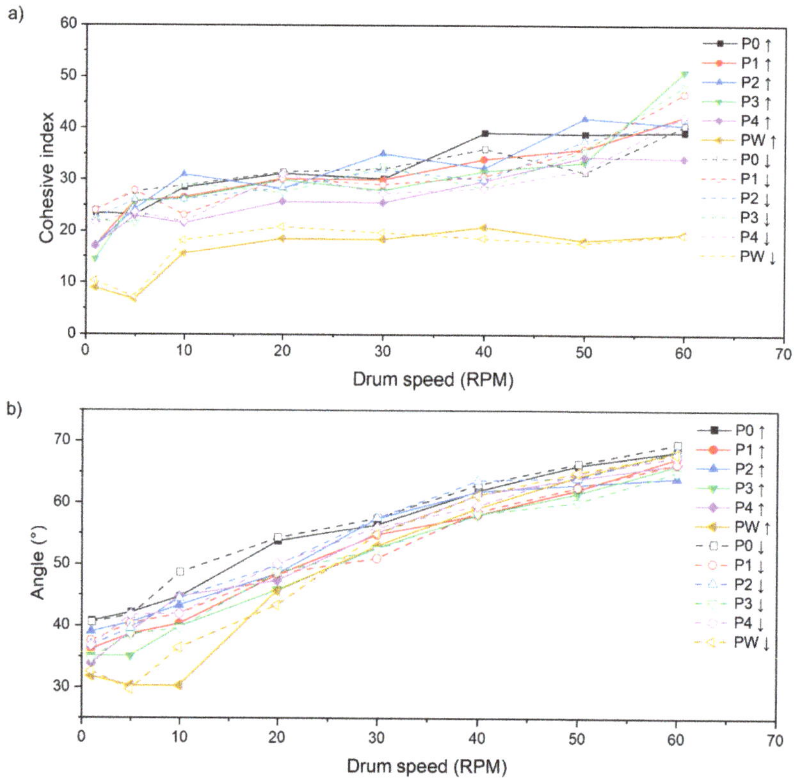

Figure 6. (**a**) The cohesion index for the AlSi7Mg0.6 powder in different processing states regarding the rotational speed measuring drum; (**b**) Mean avalanche angle values for the AlSi7Mg0.6 powders in various states with regards to the rotational speed of the measuring drum.

3.1.3. Physicochemical Properties

The absorption measured in the range of 1020–1100 nm of the wavelength is comparable, and in the case of P0–P4 powders is between 57 and 62%. Waste powder PW exhibits a 25% higher absorption and reaches a value of ~78% (Figure 7a). A slight difference in the sample after the first manufacturing process (P1) can be distinguished by zooming in the plot to a narrower absorption scale (Figure 7b). It can be seen that the registered signal in the entire wavelength range is about 5% higher than the rest of the analyzed samples.

A polynomial curve fitting of the laser absorption measurement with a 95% confidence interval confirms the observed difference between P1 and P0, P2, P3, and P4 samples.

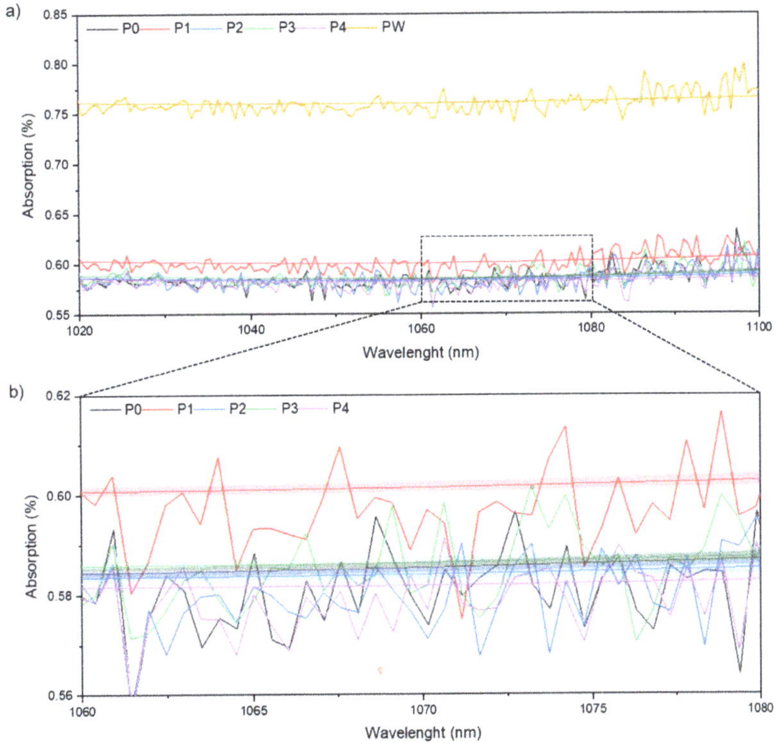

Figure 7. Laser light absorption for wavelengths from 1020 to 1110 nm of AlSi7Mg0.6 powder and polynomial curve fitting with 95% confidence interval (**a**) P0, P1, P2, P3, and P4 state and PW absorption for wavelength from 1020 to 1110 nm for (**b**) magnification for the wavelength from 1060 to 1080 nm and for P0, P1, P2, P3, and P4 states.

3.1.4. Chemical and Phase Composition

In order to determine the influence of the powder degradation on the functional properties of the samples obtained by the LPBF method, the chemical composition analysis using the XRF method was performed. Table 4 and Figure 8 show the analysis results for the bulk samples (S0 ... S4) and the powders (P0 ... P4).

The chemical composition remains unchanged. The minor discrepancies are within the error limits and the XRF method accuracy. In the case of aluminium alloys, the evaporation of some elements (for example, Mg or Zn) is one of the critical aspects of LPBF processing [28]. In the present study, it can be noticed that the magnesium content in the alloy did not change significantly in the case of both the powder and solid samples.

Table 4. Chemical composition of powder (in different states) and samples manufactured from each powder in wt.%.

Composition	Al	Si	Mg	Fe	Ti	Cu	Zn	Other Each	Other Total
AlSi7Mg0.6—SLM Solutions—materials datasheet	Bal.	6.50–7.50	0.45–0.70	0.19	0.25	0.05	0.07	0.03	0.10
			Powder specimens						
P0		6.13	0.64	0.09	0.08	0.001	0.010	-	-
P1		6.13	0.72	0.09	0.085	0.001	0.007	-	-
P2	Bal.	5.96	0.61	0.11	0.092	0.001	0.001	-	-
P3		6.04	0.62	0.12	0.098	0.001	0.001	-	-
P4		6.11	0.65	0.10	0.082	0.001	0.001	-	-
			LPBF specimens						
S0		6.20	0.729	0.052	0.056	0.0006	0.004	-	-
S1		6.188	0.715	0.049	0.052	0.0006	0.004	-	-
S2	Bal.	6.25	0.727	0.049	0.053	0.0007	0.004	-	-
S3		6.288	0.735	0.052	0.056	0.0007	0.004	-	-
S4		6.220	0.729	0.051	0.057	0.0006	0.004	-	

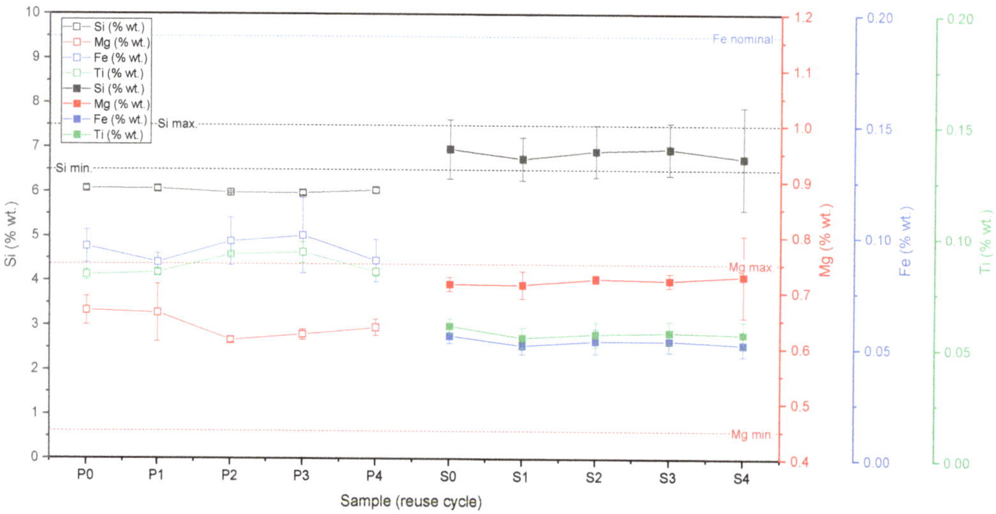

Figure 8. Chemical composition of powder (in different state—P0 . . . P4) and bulk samples (S0 . . . S4) from AlSi7Mg0.6 in wt.% measured by XRF method.

3.2. *Sample Characterization*

3.2.1. Porosity

The XCT analyses aimed to assess the internal structure of the samples manufactured from successive iterations of the reused powder and to check the impact of powder degradation on the defects' formation. For this purpose, six samples from each series were scanned, as presented in Figure 9.

The porosity analysis was performed in the same method for each sample for a region of interest (ROI) 25 mm high (Figure 9a), corresponding to the gauge section of the tensile sample (Figure 3). The results of the porosity analyses are presented in the form of graphs of the mean values of (a) porosity, (b) maximum pore diameter, and (c) maximum pore volume, taking into account the standard deviation of the results (Figure 10).

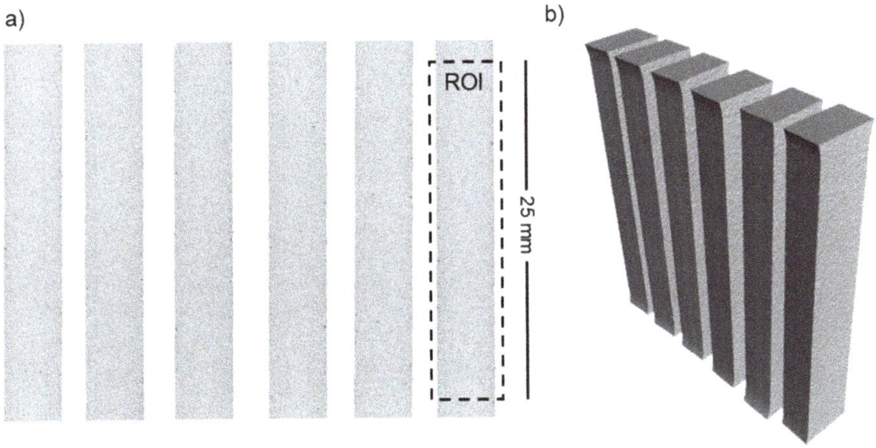

Figure 9. Two-dimensional cross-section through the reconstructed sample (**a**) and a three-dimensional view of the obtained models (**b**). The reconstruction looks similar for all series. Presented reconstruction refers to series S0.

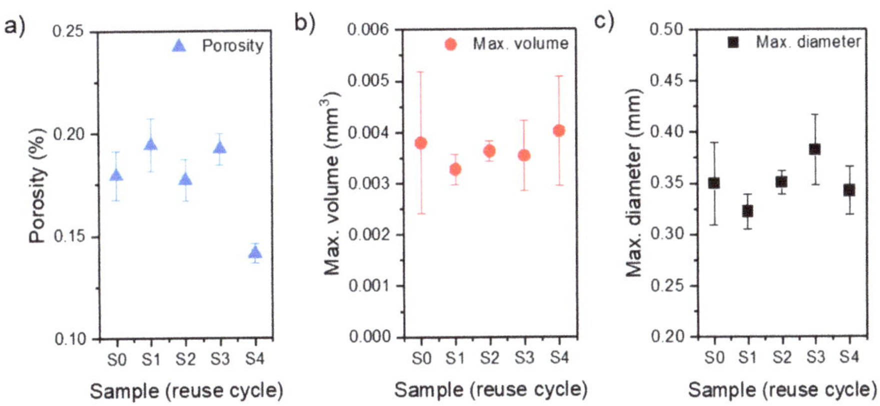

Figure 10. Analysis of the porosity of the measuring part of the samples (error—95% confidence interval): (**a**) a graph of measured porosity (XCT), (**b**) a graph of the maximum pore diameter, and (**c**) a graph of the maximum pore volume.

Based on the obtained results, there is no noticeable trend of changes in the porosity of the samples concerning subsequent iterations of the input material processing. The exception is series 4four for which the recorded porosity values are the lowest. However, the results are evenly distributed for all series concerning the mean value of the maximum diameter size and pore volume.

The diameters of the registered pores and their shape (sphericity) were evaluated for three samples showing the highest porosity from each series. In Figure 11a, box plots of pore diameters are presented. In Figure 11b, box plots of pore sphericities are presented. Pore sphericity is defined as the aerial ratio of the sphere to the pore where the sphere outlines the pore. The closer the value to one, the higher the pore sphericity is [29].

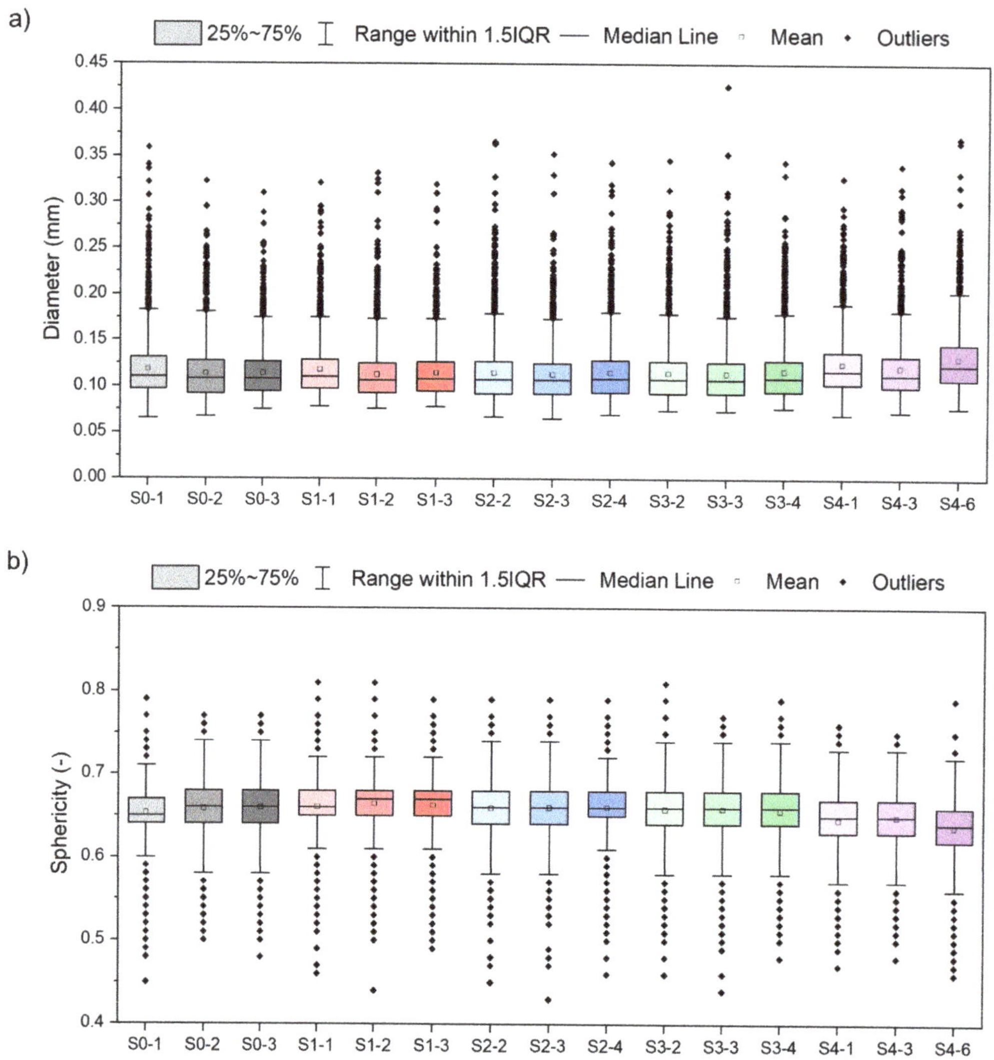

Figure 11. Box plots of pore diameter (**a**) and pore sphericity (**b**) based on CT data for three of the most porous samples of each series: S0 to S4; outliners qualified using the 1.5IQR method (IQR—interquartile range).

The results of the data from the individual series and between all the processes coincide with each other. There is a slight difference in the mean pore sizes for all S4 samples compared to the rest of the series (S0–S3). There is also a slight decrease in mean sphericity. Nevertheless, pore diameter and sphericity show significant deviation, here expressed as 1.5 times the interquartile range (1.5IQR). Additionally, each series and sample show outliners. A low number of pores in each series show larger diameters than 1.5IQR. A low number of pores show also smaller and larger sphericities than 1.5IQR.

The total number of pores recorded in the ROI volume distinguishes the S0–S4 series and the S4 series. The pore number is about 30% smaller for the S4 series than for the S0–S3 series. A comparison of the pore morphology and pore count is presented in Figure 12.

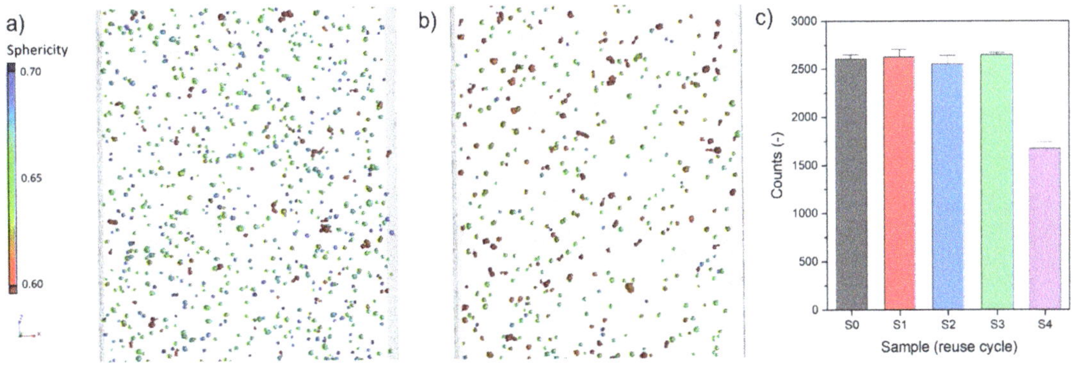

Figure 12. Pore reconstruction recorded for samples (**a**) S1-1 and (**b**) S4-6; (**c**) Pore count in ROI for each respective series (S0–S4)—histogram based on samples analysed in Figure 11 (error—95% confidence interval).

The samples' pore distribution homogeneity was compared for the XZ, YZ, and ZX (Figure 13a–e) and samples S0-1 and S4-1. The porosity analysis was performed for sequential ROIs with heights equal to 0.25 mm [25]. The most significant changes in porosity were noted for the XY plane. It results from the presence of subsurface pores, which are a defect caused by the selected LPBF boundary process parameters and strategy. Hence, the porosity in this plane has the highest values, even up to 1.35% (Figure 13f). It is worth noting that this phenomenon does not occur in the other planes of the analysis due to the machining and ROI selection. Hence, the global values do not exceed 0.22%, as shown in Figure 13g,h.

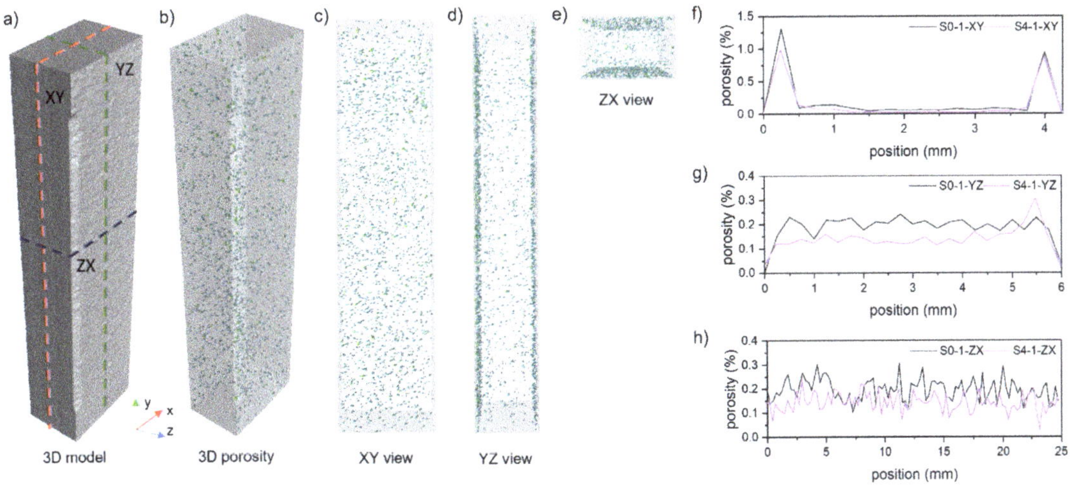

Figure 13. Pore distribution in the sample S1-1 (**b**), visible in planes (**a**), XY (**c**), YZ (**d**), ZX (**e**), and graphs of porosity in the analysed planes (**f–h**).

3.2.2. Microstructure

The microstructure of the samples is typical for the additively manufactured aluminium alloys and is characterized by a fine, columnar–dendritic structure. The CLSM (Figure 14) and SEM (Figure 15) microscopic images do not differ between each series. Figure 14a–e represent the XZ plane of specimens, and Figure 14f shows the XY plane of an

S0 specimen to show the characteristic fusion lines and material texture in planes parallel (XZ) and perpendicular (XY) to the LPBF build direction (BD).

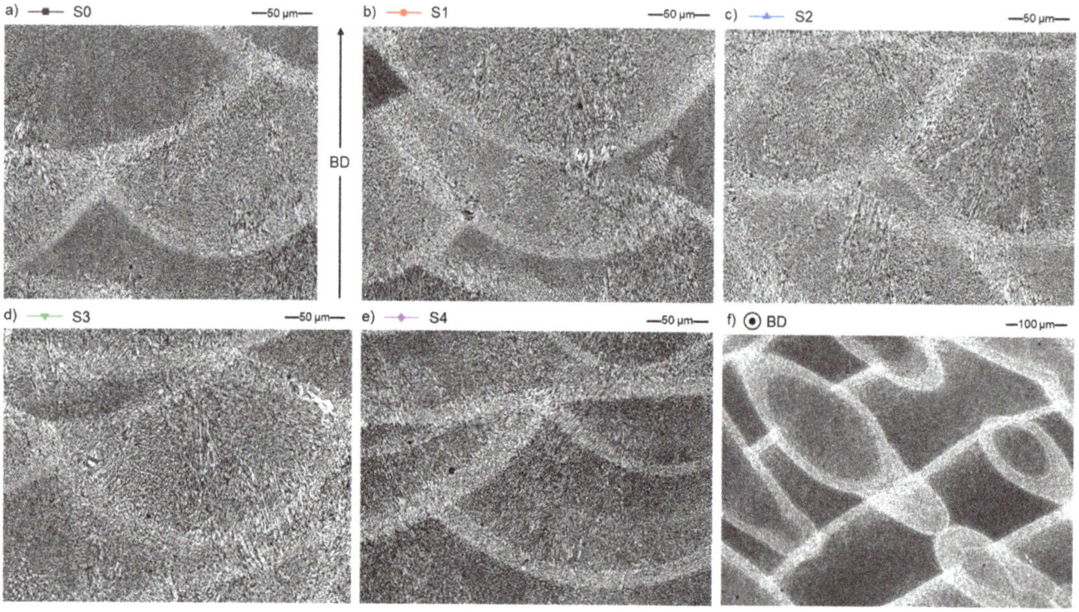

Figure 14. General microstructure of AlSi7Mg0.6 samples cross-section parallel to build direction (BD) for S0–S4 and perpendicular for S0, CLSM; (**a**) S0; (**b**) S1; (**c**) S2; (**d**) S3; (**e**) S4; and (**f**) S0.

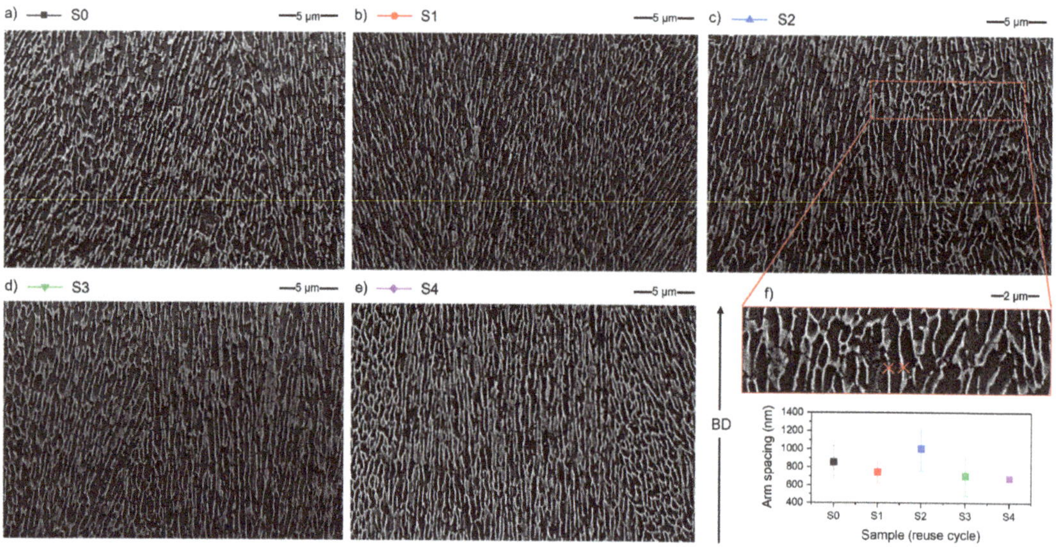

Figure 15. Microstructure of AlSi7Mg0.6 samples, SEM/BSE; (**a**) S0; (**b**) S1; (**c**) S2; (**d**) S3; (**e**) S4; and (**f**) Dendrite arm spacing measurement example and a graph of dendrite arm spacing—avg. from 6 random measurements for each sample from series S0 to S4.

The microstructure of AlSi7Mg0.6 is typical for hypoeutectic alloy [30] (Figure 15a–e). It consists of the α-Al phase (grey background on SEM microphotographs) and a network of inter-dendritic regions rich in Si (light grey colour).

On SEM micrographs, the dendrite arm spacing was measured (Figure 15f) to assure no difference and no influence of powder reuse on the microstructure features. The results of measurements are presented in Figure 15f, and there are minor differences within samples, but all of the measurement differences fall within the error limits. Therefore, it can be concluded that there are no differences, especially considering that such a measurement will be sensitive to the collection site and the orientation of the specimen cross-section to the examined dendritic structure.

3.3. Mechanical Properties

Performed static tensile tests showed no visible powder degradation trend due to its reuse. All series show a high UTS of 390–400 MPa. The highest mean values were obtained for samples from series S0, S1, and S2, with narrow confidence intervals (Figure 16a) and UTSs above 395 MPa. The series S3 and S4 have lower mean values (below 395 MPa), but the confidence intervals reach the mean value of the rest of the analysed series.

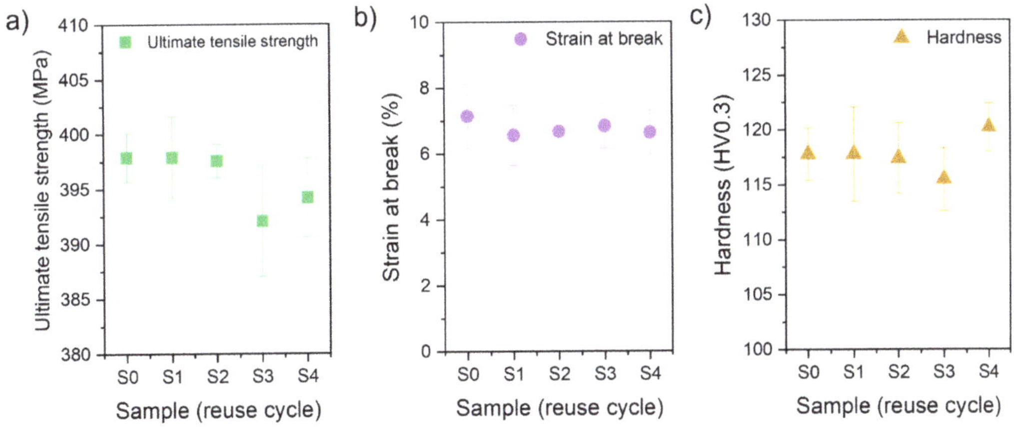

Figure 16. Mechanical properties of LPBF AlSi7Mg0.6 samples S0–S4 (error—95% confidence interval): (**a**) Ultimate tensile strength, (**b**) Strain at break, and (**c**) Hardness (HV0.3).

The strain at break values shows high consistency and each confidence interval overlaps. Each series show 6–7% of strain at break (Figure 16b).

The hardness of the samples (mean for all samples of 118 ± 3 HV0.3) is comparable with the value from a material datasheet (112 ± 3 HV10) [21]. Furthermore, each confidence interval overlaps. The summary for the hardness measurements for each series of samples (S0–S4) is presented in Figure 16c.

The obtained mechanical properties are in line with properties of AlSi7Mg0.6 alloy processed by LPBF found in the literature (Table 5). Despite the slight differences in UTS, hardness, and strain at break, the typical correlation is maintained. A lower hardness results in a higher elongation and a lower UTS. The phenomena are related to the parameters of the LPBF process and resulting solidification. Faster cooling creates finer, less ductile microstructures, thus producing higher UTS [31].

Table 5. Mechanical properties of LPBF AlSi7Mg0.6 specimens produced in XZ direction (comparison of different sources).

Property	This Research *	SLM Solutions [21]	Pereira et al. [30]
UTS, MPa	398 ± 2	375 ± 17	435 ± 18
Strain at break, %	7 ± 1	8 ± 2	3 ± 1
Hardness, HV	119 ± 2.5	112 ± 3	136.4 ± 2.5

* Presented values concerning series 0 (virgin powder).

4. Discussion

The hypothesis based on the literature data about other powder materials [32–34] was that AlSi7Mg0.6 powder would degrade with each cycle: its surface would be oxidised and PSD would increase toward bigger particles [35]. Therefore, it was expected that the laser absorption would change due to the surface oxidation and PSD, impacting the process conditions and the final sample properties. Furthermore, the literature claimed that such surface oxidation is typical for highly reactive materials such as titanium [6], nickel [10,36], and aluminium alloys [37]. However, the results shown above demonstrate that the analysed AlSi7Mg0.6 powder is highly stable in terms of the laser absorption level during five consecutive LPBF processes.

The minimal absorption increase for the P1 powder sample can be connected to multiple reflections. According to [38], it can appear when the beam is reflected away by the powder bed more than one time. Authors of [38] mention that the reason for multiple reflections is the grain size and shape. Both the particles smaller than the beam's diameter and non-spherical particles can cause the multiple reflections and therefore increased the absorption. Therefore, the measurement results could be influenced by the powder layer composition during the measurement or the place of powder sampling.

Nevertheless, if the ranges of the laser absorption values are compared and not the values of the averaged polynomial curve fitting (Figure 7), the difference in the series P1 is less significant. Besides, it should be mentioned that the surface structure significantly impacts the absorption level [39]. Therefore, a significantly higher absorption of the waste powder (PW) confirms a surface structure and PSD influence on the laser absorption level.

The other characteristic is powder morphology. Two distinctive features of the powder set it apart from the virgin powders reported in the powder reuse literature. Therefore, it should be considered:

(1) Small powder particles (in the form of satellites and loose particles) are usually found in virgin powders [10,15]. Such powders during reuse are losing small particles. Therefore, the changing PSD translates into the change of powder flow or laser absorption [10]. The powder analysed in this work does not have many small powder particles. Moreover, as mentioned in Section 2.1 (materials and processing), virgin powder was pre-sieved before the first P0 process. That is why there is no significant difference in PSD analysis. The sieving procedure between processes successfully separate agglomerations and partial melted particles, which can impact the process.

(2) The shape of particles. The analysed powder has elongated, potato-like shape particles. However, it shows an acceptable level of flowability and processability. The multiple processing of AlSi7Mg0.6 powder in LPBF does not change its flow properties compared to other materials such as titanium [40] or Inconel 718 [10] powder. According to the [41], the flowability can be even improved between 6 and 15 cycles of reuse.

In terms of chemical composition, there were two expectations or hypotheses. The first one is about a general change of chemical composition due to multiple powder reuse. The main difference considered is the change of zinc. However, this change is minor and does not affect the properties of material.

The second one was about magnesium evaporation [28]. As a result of five consequent processes, magnesium's evaporation was not detected in powder or bulk samples.

According to [42], the small addition (low content) of magnesium (up to 1.5%) positively influences the microstructure and processability of Al-Si alloys. It should be emphasised that magnesium in low-magnesium Al-Si alloys tends to condense and be studded at the cell boundaries, especially at the nodes of cell boundaries [42]. Therefore, the results described above are well-aligned with those in the literature.

However, magnesium evaporation is the main problem in AM of high-magnesium Al-Mg alloys, but is not confirmed by the literature for the Al-Si alloys except for one publication [43], where authors describe the evaporation of magnesium and zinc for different aluminium alloys. Process parameters are possible reasons for the difference between the results published in [43] and the presented research. High-power LPBF processing should contribute to this effect. An interconnection between magnesium content, densification, and applied energy density affects the influence of process parameters on magnesium evaporation. In the case of magnesium content <2.0%, there is no need to use high-energy density to densify samples [42]; therefore, the risk of magnesium evaporation is lower.

As for the powder samples, the analysed AlSi7Mg0.6 bulk specimens (S0–S4) do not show significant proofs of powder degradation. All bulk specimens (S0–S4) show comparable and repeatable microstructures (similar texture and dendrite arm spacing), which are typical for AM-processed hypoeutectic aluminium alloys.

A similar consistence of results is found within mechanical properties and hardness. The difference between each series (S0–S4) is lower than the standard deviation of the results. Even if the error is neglected, the determined mean values are within 5 MPa for the UTS, 0.5 p.p. for the break at strain, and 5 HV0.3 for the hardness.

In the paper [44], the influence of powder reuse on the mechanical properties of AlSi10Mg alloy within eight consecutive LPBF processes without rejuvenation is presented. The study shows that significant (visible) changes can appeared after the 5–6 LPBF processes. However, in the discussed paper, error limits for each series are not presented. If the changes can be fitted into the error limits as in this work, therefore AlSi10Mg powder degradation shown in [44] could be minor.

On the contrary, if results from [44] are considered, series S4 may be the critical point, after which some changes could appear. Even if all powder samples (P0–P4) and bulk specimens (S0–S4) show repeatable properties with changes within error limits, certain signs could indicate some initial degradation. A small decrease in dendrite arm space can be observed, a small change in chemical composition and a higher mean hardness. All the above-mentioned minor changes can be translated into the change of detected bulk samples' porosity distribution. The pore count is approximately 30% lower for the S4 series than for the S0–S3 series.

The authors of [4] showed in their work the difference between two different powder reuse strategies. The strategy used in the presented work (continuous reuse/single batch) has its limit regarding powder availability for producing the subsequent samples. At a certain moment, it is impossible to process samples with the same geometry due to the lack of powder. Therefore, the number of cycles is limited by the quantity of powder without rejuvenation.

The approach with frequent refreshing from one point is more similar to real production conditions. However, at the same time, the powder degradation during the following cycles is levelled/slowed down by the constant addition of virgin powder.

In the presented study, an attempt was made to maintain identical LPBF processing conditions at each stage of collective ageing powder reuse. Each of the LPBF processes was carried out using the same parameters, samples with a constant cross-section were fabricated, and the conditions of the LPBF process were strictly controlled (platform temperature, pressure in the chamber, gas flow speed, oxygen level, laser beam power variation, etc.). The variability of the parameters recorded during LPBF processes did not exceed 5%. In addition, the powder after each step was screened twice to ensure the adequate separation of oversized particles. The experiment was stopped after five cycles as the amount of powder that remained in circulation was insufficient to allow the fabrication of full height tensile samples. During the process, approximately 30% of the initial amount of powder (15 kg) was used to fabricate samples (4.8 kg) and 0.3 kg (2%) was screened as waste. Given the above, the analysed AlSi7Mg0.6 powder showed a high stability during reuse in the LPBF process. After the five consequent processes run without adding fresh (virgin) powder, it is demonstrated that most properties of both powder and bulk samples remain unchanged.

Therefore, in the case of the AlSi7Mg0.6 alloy, these are conditions for which the collective ageing powder reuse strategies should be safe and repeatable. The present study's future scope is to analyse the limit of safety of AlSi7Mg0.6 powder reuse. So far, the five cycles of continuous reuse do not influence the quality of produced samples. It will be essential to design the experiment looking for those limits and create the roadmap for the first signs of powder degradation.

5. Conclusions

The presented study investigated the possibility of continuous reuse (collective ageing strategy) of AlSi7Mg0.6 powder in the laser powder bed fusion process. The five batches of AlSi7Mg0.6 powder (P0–P4) and five bulk LPBF samples (S0–S4) series were characterised for powder morphology, chemical composition, porosity, and microstructure. In addition, the mechanical properties of the LPBF AlSi7Mg0.6 specimens fabricated with reused powder were investigated to ensure the comparable properties of each reuse cycle. Based on the presented results, the following conclusions can be drawn:

1. The average particle size, morphology, and chemical composition of the virgin and continuously reused AlSi7Mg0.6 powders are comparable. The main outliner is waste powder, screened during double-sieving, which differs in each property from the virgin and continuously reused power.
2. Mechanical properties of the LPBF AlSi7Mg0.6 samples manufactured using continuous reused powder are similar to the LPBF AlSi7Mg0.6 alloy manufactured samples using virgin powder. It confirms that the approach of continuously reused powder can be reasonably used in the LPBF process without a negative effect on the quality of the final product.
3. The collective ageing powder reuse strategy is considered to give repeatable LPBF process results and is recommended for the AlSi7Mg0.6 alloy within at least five consecutive LPBF processes.
4. The presented findings should be only considered when: LPBF process parameters are strictly controlled; the powder is double-sieved in each process; the virgin powder shows a similar morphology to the powder used in this study—it is free from small powder particles and is pre-sieved before use.
5. Samples manufactured within the fifth reuse cycle (series P4, S4) showed signs indicating initial degradation. These changes, however, mostly fit within error limits. Further studies should be looking at the high-cycle reuse of AlSi7Mg0.6 alloy in LPBF to set the reuse limit and create the roadmap for the first signs of powder degradation.

Author Contributions: Conceptualization, I.S. and K.G.; methodology, I.S., K.G. and G.Z.; software, E.G., G.Z. and K.G.; validation, A.P., K.G. and G.Z.; formal analysis, R.R. and A.P.; investigation, I.S., E.G., G.Z., K.G. and D.S.; resources, T.K., D.S., K.K., R.R. and M.Z.; data curation, K.G., E.G. and D.S.; writing—original draft preparation, I.S., K.G., A.P., E.G. and G.Z.; writing—review and editing, K.G., R.R., E.G., K.K., T.K. and M.Z.; visualization, K.G., E.G. and K.K.; supervision, T.K. and M.Z.; project administration, I.S.; funding acquisition, M.Z. All authors have read and agreed to the published version of the manuscript.

Funding: This work was supported by BTU Graduate Research School (BTU Flagship Fellowship Programme).

Institutional Review Board Statement: Not applicable.

Informed Consent Statement: Not applicable.

Data Availability Statement: Not applicable.

Conflicts of Interest: The authors declare no conflict of interest. The funders had no role in the design of the study; in the collection, analyses, or interpretation of data; in the writing of the manuscript, or in the decision to publish the results.

References

1. Ngo, T.D.; Kashani, A.; Imbalzano, G.; Nguyen, K.T.Q.; Hui, D. Additive manufacturing (3D printing): A review of materials, methods, applications and challenges. *Compos. Part B Eng.* **2018**, *143*, 172–196. [CrossRef]
2. Javaid, M.; Haleem, A.; Singh, R.P.; Suman, R.; Rab, S. Role of additive manufacturing applications towards environmental sustainability. *Adv. Ind. Eng. Polym. Res.* **2021**, *4*, 312–322. [CrossRef]
3. *ASTM ISO/ASTM 52900-21*; Additive Manufacturing—General Principles—Fundamentals and Vocabulary. ASTM International: West Conshohocken, PA, USA, 2021. [CrossRef]
4. Moghimian, P.; Poirié, T.; Habibnejad-Korayem, M.; Zavala, J.A.; Kroeger, J.; Marion, F.; Larouche, F. Metal powders in additive manufacturing: A review on reusability and recyclability of common titanium, nickel and aluminum alloys. *Addit. Manuf.* **2021**, *43*, 102017. [CrossRef]
5. Simchi, A. The role of particle size on the laser sintering of iron powder. *Metall. Mater. Trans. B* **2004**, *35*, 937–948. [CrossRef]
6. Derimow, N.; Hrabe, N. Oxidation in Reused Powder Bed Fusion Additive Manufacturing Ti-6Al-4V Feedstock: A Brief Review. *JOM* **2021**, *73*, 3618–3638. [CrossRef]
7. Leung, C.L.A.; Marussi, S.; Towrie, M.; Atwood, R.C.; Withers, P.J.; Lee, P.D. The effect of powder oxidation on defect formation in laser additive manufacturing. *Acta Mater.* **2019**, *166*, 294–305. [CrossRef]
8. Hryha, E.; Shvab, R.; Gruber, H.; Leicht, A.; Nyborg, L. Surface Oxide State on Metal Powder and its Changes during Additive Manufacturing: An Overview. *La Metall. Ital.* **2018**, *3*, 34–39.
9. Santecchia, E.; Mengucci, P.; Gatto, A.; Bassoli, E.; Defanti, S.; Barucca, G. Cross-Contamination Quantification in Powders for Additive Manufacturing: A Study on Ti-6Al-4V and Maraging Steel. *Materials* **2019**, *12*, 2342. [CrossRef]
10. Gruber, K.; Smolina, I.; Kasprowicz, M.; Kurzynowski, T. Evaluation of Inconel 718 Metallic Powder to Optimize the Reuse of Powder and to Improve the Performance and Sustainability of the Laser Powder Bed Fusion (LPBF) Process. *Materials* **2021**, *14*, 1538. [CrossRef]
11. Lutter-Günther, M.; Gebbe, C.; Kamps, T.; Seidel, C.; Reinhart, G. Powder recycling in laser beam melting: Strategies, consumption modeling and influence on resource efficiency. *Prod. Eng.* **2018**, *12*, 377–389. [CrossRef]
12. Heiden, M.J.; Deibler, L.A.; Rodelas, J.M.; Koepke, J.R.; Tung, D.J.; Saiz, D.J.; Jared, B.H. Metal Powder Feedstock Reuse in Additive Manufacturing: Characterization of 316L Stainless Steel. *Addit. Manuf.* **2018**, *25*, 84–103.
13. Alamos, F.J.; Schiltz, J.; Kozlovsky, K.; Attardo, R.; Tomonto, C.; Pelletiers, T.; Schmid, S.R. Effect of powder reuse on mechanical properties of Ti-6Al-4V produced through selective laser melting. *Int. J. Refract. Met. Hard Mater.* **2020**, *91*, 105273. [CrossRef]
14. Fiegl, T.; Franke, M.; Raza, A.; Hryha, E.; Körner, C. Effect of AlSi10Mg0.4 long-term reused powder in PBF-LB/M on the mechanical properties. *Mater. Des.* **2021**, *212*, 110176. [CrossRef]
15. Cordova, L.; Bor, T.; de Smit, M.; Carmignato, S.; Campos, M.; Tinga, T. Effects of powder reuse on the microstructure and mechanical behaviour of Al–Mg–Sc–Zr alloy processed by laser powder bed fusion (LPBF). *Addit. Manuf.* **2020**, *36*, 101625. [CrossRef]
16. Da Silva, A.; Belelli, F.; Lupi, G.; Bruzzo, F.; Brandau, B.; Maier, L.; Pesl, A.; Frostevarg, J.; Casati, R.; Lopez, E.; et al. Influence of aluminium powder aging on Directed Energy deposition. *Mater. Des.* **2022**, *218*, 110677. [CrossRef]
17. Mostow, N.; Diegel, O.; Wohlers, T. Findings from Wohlers Report 2022: Taking a chance on new technologies and the evolving materials mix. *Met. Addit. Manuf.* **2022**, *8*, 153–156.
18. Armstrong, K.O.; Price, C.; Su, J.; Wang, A.; Post, B.; Chesser, P.; Polsky, Y. *Study of Additive Manufacturing Applications to Geothermal Technologies Final Project Report*; Oak Ridge National Lab. (ORNL): Oak Ridge, TN, USA, 2021. [CrossRef]
19. Lorusso, M.; Trevisan, F.; Calignano, F.; Lombardi, M.; Manfredi, D. A357 Alloy by LPBF for Industry Applications. *Materail* **2020**, *13*, 1488. [CrossRef]
20. Baier, M.; Sinico, M.; Witvrouw, A.; Dewulf, W.; Carmignato, S. A novel tomographic characterisation approach for sag and dross defects in metal additively manufactured channels. *Addit. Manuf.* **2021**, *39*, 101892. [CrossRef]
21. Al-Alloy AlSi7Mg0,6/EN AC-42200; Material Data Sheet, SLM Solutions; Lubeck, Germany. Available online: https://www.slm-solutions.com/fileadmin/Content/Powder/MDS/MDS_Al-Alloy_AlSi7Mg0_6_0219_EN.pdf (accessed on 18 July 2022).
22. *ASTM F3456-22*; Standard Guide for Powder Reuse Schema in Powder Bed Fusion Processes for Medical Applications for Additive Manufacturing Feedstock Materials. ASTM International: West Conshohocken, PA, USA, 2022. [CrossRef]
23. *ASTM E8/E8M*; Standard Test Methods for Tension Testing of Metallic Materials. ASTM International: West Conshohocken, PA, USA, 2016. [CrossRef]
24. Ziółkowski, G.; Chlebus, E.; Szymczyk, P.; Kurzac, J. Application of X-ray CT method for discontinuity and porosity detection in 316L stainless steel parts produced with SLM technology. *Arch. Civ. Mech. Eng.* **2014**, *14*, 608–614. [CrossRef]
25. Ziółkowski, G.; Gruber, K.; Tokarczyk, E.; Roszak, R.; Ziegenhorn, M. X-ray Computed Tomography for the ex-situ mechanical testing and simulation of additively manufactured IN718 samples. *Addit. Manuf.* **2021**, *45*, 102070. [CrossRef]

26. Szymczyk, P.; Hoppe, V.; Ziółkowski, G.; Smolnicki, M.; Madeja, M. The effect of geometry on mechanical properties of Ti6Al4V ELI scaffolds manufactured using additive manufacturing technology. *Arch. Civ. Mech. Eng.* **2020**, *20*, 11. [CrossRef]
27. Flowability and Cohesion Determination of Metal Powders using Granuheap and Granudrum. Available online: https://www.granutools.com/en/news/84_flowability-and-cohesion-determination-of-metal-powders-using-granuheap-and-granudrum (accessed on 28 June 2022).
28. Rao, H.; Giet, S.; Yang, K.; Wu, X.; Davies, C.H.J. The influence of processing parameters on aluminium alloy A357 manufactured by Selective Laser Melting. *Mater. Des.* **2016**, *109*, 334–346. [CrossRef]
29. Ziółkowski, G.; Grochowska, E.; Kęszycki, D.; Gruber, P.; Hoppe, V.; Szymczyk-Ziółkowska, P.; Kurzynowski, T. Investigation of porosity behavior in SLS polyamide-12 samples using ex-situ X-ray computed tomography. *Mater. Sci. Pol.* **2021**, *39*, 436–445. [CrossRef]
30. Pereira, J.C.; Gil, E.; Solaberrieta, L.; San Sebastián, M.; Bilbao, Y.; Rodríguez, P.P. Comparison of AlSi7Mg0.6 alloy obtained by selective laser melting and investment casting processes: Microstructure and mechanical properties in as-built/as-cast and heat-treated conditions. *Mater. Sci. Eng. A* **2020**, *778*, 139124. [CrossRef]
31. Gruber, K.; Stopyra, W.; Kobiela, K.; Madejski, B.; Malicki, M.; Kurzynowski, T. Mechanical properties of Inconel 718 additively manufactured by laser powder bed fusion after industrial high-temperature heat treatment. *J. Manuf. Process.* **2022**, *73*, 642–659. [CrossRef]
32. Heiden, M.J.; Deibler, L.A.; Rodelas, J.M.; Koepke, J.R.; Tung, D.J.; Saiz, D.J.; Jared, B.H. Evolution of 316L stainless steel feedstock due to laser powder bed fusion process. *Addit. Manuf.* **2019**, *25*, 84–103. [CrossRef]
33. Shanbhag, G.; Vlasea, M. Powder Reuse Cycles in Electron Beam Powder Bed Fusion—Variation of Powder Characteristics. *Materail* **2021**, *14*, 4602. [CrossRef]
34. Cordova, L.; Campos, M.; Tinga, T. Revealing the Effects of Powder Reuse for Selective Laser Melting by Powder Characterization. *JOM* **2019**, *71*, 1062–1072. [CrossRef]
35. Yang, X.; Gao, F.; Tang, F.; Hao, X.; Li, Z. Effect of Surface Oxides on the Melting and Solidification of 316L Stainless Steel Powder for Additive Manufacturing. *Metall. Mater. Trans. A Phys. Metall. Mater. Sci.* **2021**, *52*, 4518–4532. [CrossRef]
36. Gasper, A.N.D.; Szost, B.; Wang, X.; Johns, D.; Sharma, S.; Clare, A.T.; Ashcroft, I.A. Spatter and oxide formation in laser powder bed fusion of Inconel 718. *Addit. Manuf.* **2018**, *24*, 446–456. [CrossRef]
37. Hauser, T.; Reisch, R.T.; Breese, P.P.; Nalam, Y.; Joshi, K.S.; Bela, K.; Kamps, T.; Volpp, J.; Kaplan, A.F.H. Oxidation in wire arc additive manufacturing of aluminium alloys. *Addit. Manuf.* **2021**, *41*, 101958. [CrossRef]
38. Brandau, B.; Da Silva, A.; Brueckner, F.; Kaplan, A.F.H. Absorbance study of powder conditions for laser additive manufacturing. *Mater. Des.* **2022**, *216*, 110591. [CrossRef]
39. Zhou, Y.H.; Zhang, Z.H.; Wang, Y.P.; Liu, G.; Zhou, S.Y.; Li, Y.L.; Shen, J.; Yan, M. Selective laser melting of typical metallic materials: An effective process prediction model developed by energy absorption and consumption analysis. *Addit. Manuf.* **2019**, *25*, 204–217. [CrossRef]
40. Harkin, R.; Wu, H.; Nikam, S.; Yin, S.; Lupoi, R.; McKay, W.; Walls, P.; Quinn, J.; McFadden, S. Powder Reuse in Laser-Based Powder Bed Fusion of Ti6Al4V—Changes in Mechanical Properties during a Powder Top-Up Regime. *Materials* **2022**, *15*, 2238. [CrossRef]
41. Powell, D.; Rennie, A.; Geekie, L.; Burns, N. Understanding powder degradation in metal additive manufacturing to allow the upcycling of recycled powders. *J. Clean. Prod.* **2020**, *268*, 122077. [CrossRef]
42. Kimura, T.; Nakamoto, T.; Ozaki, T.; Sugita, K.; Mizuno, M.; Araki, H. Microstructural formation and characterization mechanisms of selective laser melted Al–Si–Mg alloys with increasing magnesium content. *Mater. Sci. Eng. A* **2019**, *754*, 786–798. [CrossRef]
43. Mauduit, A.; Pillot, S.; Gransac, H. Study of the suitability of aluminum alloys for additive manufacturing by laser powder bed fusion. *UPB Sci. Bull. Ser. B Chem. Mater. Sci.* **2017**, *79*, 219–238.
44. Del Re, F.; Contaldi, V.; Astarita, A.; Palumbo, B.; Squillace, A.; Corrado, P.; Di Petta, P. Statistical approach for assessing the effect of powder reuse on the final quality of AlSi10Mg parts produced by laser powder bed fusion additive manufacturing. *Int. J. Adv. Manuf. Technol.* **2018**, *97*, 2231–2240. [CrossRef]

materials

MDPI

Article

Numerical Simulation on Thermal Stresses and Solidification Microstructure for Making Fiber-Reinforced Aluminum Matrix Composites

Chenyang Xing [1,2], Reihaneh Etemadi [3], Krishna M. Pillai [3], Qian Wang [1,2] and Bo Wang [1,2,*]

[1] State Key Laboratory of Advanced Special Steel, Shanghai University, Shanghai 200072, China; xingcy136358@163.com (C.X.); wq06@shu.edu.cn (Q.W.)
[2] School of Materials Science and Engineering, Shanghai University, Shanghai 200072, China
[3] Department of Mechanical Engineering, University of Wisconsin-Milwaukee, Milwaukee, WI 5322, USA; retemadi@uwm.edu (R.E.); krishna@uwm.edu (K.M.P.)
* Correspondence: bowang@shu.edu.cn

Abstract: The fabrication of fiber-reinforced metal matrix composites (MMCs) mainly consists of two stages: infiltration and solidification, which have a significant influence on the properties of MMCs. The present study is primarily focused on the simulation of the solidification process and the effect of the active cooling of fibers with and without nickel coating for making the continuous carbon fiber-reinforced aluminum matrix composites. The thermomechanical finite element model was established to investigate the effects of different cooling conditions on the temperature profile and thermal stress distributions based on the simplified physical model. The predicted results of the temperature distribution agree well with the results of the references. Additionally, a three-dimensional cellular automata (CA) finite element (FE) model is used to simulate the microstructure evolution of the solidification process by using ProCAST software. The results show that adding a nickel coating can make the heat flux smaller in the melt, which is favorable for preventing debonding at the coating/fiber and alloy interface and obtaining a finer microstructure. In the presence of the nickel coating, the number of grains increases significantly, and the average grain size decreases, which can improve the properties of the resultant composite materials. Meanwhile, the predicting results also show that the interfaces of fiber–coating, fiber–melt, and coating–melt experience higher temperature gradients and thermal stresses. These results will lead to the phenomenon of stress concentration and interface failure. Thus, it was demonstrated that these simulation methods could be helpful for studying the solidification of fiber-reinforced MMCs and reducing the number of trial-and-error experiments.

Keywords: fiber-reinforced; MMCs; numerical simulation; temperature field; microstructure

Citation: Xing, C.; Etemadi, R.; Pillai, K.M.; Wang, Q.; Wang, B. Numerical Simulation on Thermal Stresses and Solidification Microstructure for Making Fiber-Reinforced Aluminum Matrix Composites. *Materials* **2022**, *15*, 4166. https://doi.org/10.3390/ma15124166

Academic Editors: Mostafa Hassani, Hongze Wang and Greta Lindwall

Received: 8 May 2022
Accepted: 10 June 2022
Published: 12 June 2022

1. Introduction

Metal matrix composites (MMCs) are usually composed of a metal or alloy as the continuous phase and whiskers or fibers of a reinforcing material as the second phase [1,2]. MMCs are widely used in developing materials for aerospace, electronics, and optical instruments due to their good mechanical properties, including low density, high Young's modulus, high wear, and fatigue resistance. However, factors such as poor wettability, chemical reaction at the melt–fiber interface, and larger grain size and dendrite arm spacing during solidification processing tend to restrict the development of these materials for the industry [1–8].

The pressure infiltration process (PIP) is an established technique to manufacture MMCs where liquid metal or alloy is injected into a dry porous medium called the preform, made of reinforcing fibers, and later solidified to create the solid composite. Such fabrication of MMCs includes two stages: infiltration and solidification [1,5,9,10]. The liquid–metal infiltration process is a complicated flow and transport phenomenon, which can involve the preferential flow of liquid metal through larger pores of the preform, the mechanical deformation of the preform, and the solidification of liquid metal on coming in contact with cooled fibers and surfaces [11]. Transport phenomena during infiltration govern the temperature and solute distributions at and behind the infiltration front. These phenomena are often accompanied by other phenomena such as the segregation of alloying elements and chemical reactions. Finally, the solidification of the metal matrix occurs during and after the infiltration process, resulting in the final MMC part. In practice, all these phenomena simultaneously occur during the infiltration process.

The interface problem has been a core issue in manufacturing metal matrix composites, especially for active metals such as aluminum [12–15]. High infiltration temperature is a key factor in preparing carbon and aluminum (C/Al) composites. However, at this temperature, aluminum readily reacts with carbon to form a brittle phase of Al_4C_3 between carbon fibers and the aluminum matrix, which leads to the degradation of carbon fibers and consequently results in the deterioration of the mechanical properties of the composite and leads to its early failure under load [15]. High-resolution microfractography and transmission electron microscopy show that the mechanical behavior of the carbon-fiber-reinforced Al-based matrix composites is related to the presence of brittle interfacial phases [16]. The most common way to solve this problem is to coat the surface of the carbon fibers using the vapor deposition technique with nickel or copper [15]. This coating not only reduces the reaction at the fiber–melt interface, but also improves the wettability [17–19]. By using the nanoindentation technique, A. Urena et al. [19] investigated the interfacial mechanical properties of an AA6061 composite reinforced with short carbon fibers coated with copper and nickel films. The film coating on the carbon fiber surface was applied to control the interfacial reactivity of fibers with molten aluminum during the manufacture of the composite. The results showed that the copper coating produced by electroless increases the hardness and stiffness of the aluminum matrix, and nickel coatings decrease the hardness of the matrix close to the fibers and produce a high dispersion of stiffness values, especially in the own interface and at distances above 5 μm from the fibers. Improving the interfacial bonding between fibers and melt is one of the key factors in improving the properties of the fiber-reinforced composites [15,20]. The coating can play the adhesive role on the interface, leading to an improvement in the load transfer to the fibers. It has also been observed experimentally that the presence of carbon fibers alters the microstructure of the matrix alloy created during solidification. For example, Z.G. Liu et al. [21] studied the interface in the carbon fiber-reinforced Al–Cu alloy composites. The important feature observed in their experiments was that the microstructure of the Al–Cu matrix alloy was altered due to carbon fibers.

Nickel and copper are among the widely used coating materials. Although these metallic coatings can improve the wetting of carbon materials, the formation of intermetallic compounds or carbides will reduce the mechanical properties of the composites. This shortcoming should be controlled by optimizing the coatings' thickness and the composites' fabrication parameters. In the present study, the effect on the solidification of fibers with and without nickel coating was investigated. The effect of the brittle phases or transition phases formed at interfaces is ignored in the simulation study.

The solidification of MMCs is essentially a process of nucleation and the growth of crystals for base alloys. In an actual process, the grain growth is always accompanied by the phenomena of dendrite remelting and dendrite segregation. However, this effect is often ignored in the numerical simulation of the MMC solidification due to its complexity and little effect on the overall results. Many works of the simulation on infiltration and solidification processes for making MMCs have been done, but little study has been implemented on microstructure simulations, especially the nucleation and growth of crystals during the solidification process [22–27]. The main reason is that the presence of reinforcement materials in the metal matrix composites makes the solidification process more complicated. At present, the cellular automata (CA) method and phase-field method are two commonly used methods for simulating the microstructure evolution in solidification processes [28]. Although the phase-field method is accurate for the simulation of microstructure evolution due to its foundation on thermodynamics and physically-informed parameters, its overly complex principle, the need for more enormous computational resources, and the small computational domain hinder its further industrial application. In this study, the CA method is used to simulate the nucleation and growth of crystals in the solidification processes for making MMCs.

The microstructure development in MMCs is closely related to the temperature field, which can be changed by controlling the cooling rate. Lee et al. [27] studied the effect of the cooled fibers on the solidification microstructures based on numerical simulation and experimental observation. The results show that cooling the ends of the fibers changed the cooling curves (temperature fields) to lead to the nucleation of aluminum dendrites on the surface of the fibers. In the absence of such cooling, primary aluminum nucleated away from the surface of the fibers, depositing the last freezing eutectic at the interface. A faster cooling rate will result in higher temperature gradients and the development of a fine grain structure [27,29]. In addition, the number of nucleation sites increases significantly due to the presence of the reinforcement phase, which is favorable to the formation of a large number of fine crystals. Lelito et al. [29] developed a numerical micro–macro model based on the empirical nucleation law to predict the grain density in the Mg-based MMCs. The experimental and simulation results also show that the cooling curves and matrix grain densities were a function of heat-extraction rates, mass fraction, and the particle diameter of SiC. So, one possible way for such microstructure improvement is to extend the ends of the reinforcing phase to the outside of the casting mold and cool the ends of the fibers with a heat sink. Due to a faster heat extraction, the solidification time is reduced. This method is referred to as the thermal management of fibers, which can significantly change the nature of the interface and the surrounding matrix and, therefore, the properties of the composite. Such an active fiber cooling method has also been used by researchers [30,31] to prevent damage to the nickel coating during the infiltration process. Rohatgi et al. [32] used the squeeze infiltration process to synthesize an MMC of Al-2014 reinforced with nickel-coated carbon fibers. They used a modified version of a commercial squeeze-casting machine in which the ends of the carbon fibers were made to extend out on both sides of the mold, so they were cooled due to the lower ambient temperature, resulting in a higher heat-transfer rate from the system.

Although significant work has been done to model the metal infiltration and solidification processes seen during the manufacture of MMCs, relatively less research has been conducted on modeling the evolution of grain microstructure during the solidification process [22,32]. Our study uses the fiber-based active cooling method employed in previous studies [17,30,32]. These previous works have shown that the solidification microstructures of fiber-reinforced aluminum composites can be altered by cooling the ends of the fibers extending out of the mold. Based on this method, numerical simulations have been done for a simplified model to study the temperature profile around the fibers and stress distribution with coating. However, the simulations of the microstructure evolution involving the grain nucleation and growth are not considered, which is key to the final properties of the MMCs. In this article, the temperature profile, stress distribution, and microstructure evolution around the fibers during solidification are simulated using the commercial software ProcCAST® of ESI Group. The effect of nickel coating on the solidification process is also studied. These results should be helpful in controlling and optimizing the solidification process witnessed during the making of MMCs.

2. Problem Description and Simulation Method

The conventional fiber preforms are made up of a large number of similar fiber units, as shown in Figure 1. For simplicity, a unit-cell of a cylindrical shape with a fiber and alumina melt is extracted as the calculation domain. Figure 1 presents the 3D-axisymmetric carbon fiber/aluminum model, in which the carbon fiber is located in the center and is wrapped in the aluminum melt. The physical model comes from the reference of Nguyen et al. [17,30]. This model assumes that a carbon fiber with or without coating is vertically oriented in the center and is surrounded by the Al alloy melt. Based on this model, this article discusses the temperature field and the thermal stresses and investigates the grain microstructure evolution, including the nucleation and growth process around the fiber and the trend of heat flux. The values of 1, 1, 0.2, and 0.05 units are set for radius (Ra), height (L), carbon fiber radius (Rf), and coating thickness parameters, respectively.

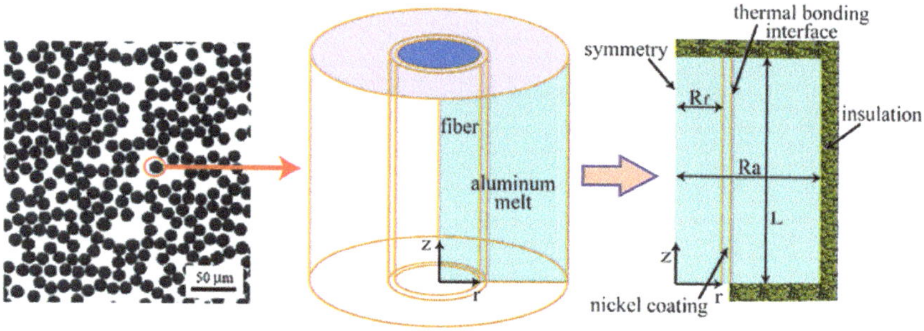

Figure 1. A schematic describing the 3D axisymmetric unit-cell for the carbon fiber/aluminum alloy MMC, as described in [17,30].

2.1. *Mathematical Model*

Considering only the postmold-fill scenario, we are going to model heat transfer, solidification, microstructure development, and stress estimation in a stationary pool of metal surrounding the fiber in the unit cell.

2.1.1. Energy Equation

The energy conservation equation is solved to study the heat transfer and solidification phenomena in the postmold-fill process for making MMCs.

$$\rho c_p \frac{\partial T}{\partial t} = \frac{\partial}{\partial x}\left(\lambda \frac{\partial T}{\partial x}\right) + \frac{\partial}{\partial y}\left(\lambda \frac{\partial T}{\partial y}\right) + \frac{\partial}{\partial z}\left(\lambda \frac{\partial T}{\partial z}\right) + Q \quad (1)$$

where ρ, c_p, t, and T are the density, the specific heat capacity, the time, and the temperature, respectively. $Q = \rho L \cdot (\partial f_s / \partial T)$, where L is the latent heat of melting and f_s is the solid fraction.

2.1.2. Thermal-Elastic-Plastic Model

The material begins to yield when the elastic deformation energy reaches a specified value under certain deformation conditions. The thermal–elastic–plastic model is adopted to simulate the deformation and thermal stress distribution during solidification for making MMCs.

The total strain increment includes thermal strain increment, elastic strain increment, and plastic strain increment. The effective stress is calculated by $\overline{\sigma}$.

The yield criterion follows the von Mises criterion:

$$\overline{\sigma} = \frac{1}{\sqrt{2}}\sqrt{(\sigma_x - \sigma_y)^2 + (\sigma_y - \sigma_z)^2 + (\sigma_z - \sigma_x)^2} \quad (2)$$

where the σ_x, σ_y, and σ_z are, respectively, the first, second, and third principal stresses.

2.1.3. Nucleation and Growth Model

A three-dimensional cellular automaton–finite element (CAFE) model is used to simulate the microstructure evolution of the Al alloy composites. The continuous nucleation model, which employs the heterogeneous nucleation approach based on Gaussian distribution, is used. It should be mentioned that heterogeneous nucleation occurs in the bulk liquid, and the surface of the fiber or nickel coating is described by two distributions of nucleation sites which became active as undercooling increases. The continuous and nondiscrete distribution function is used to describe changes in grain density, which can be determined by Gaussian distribution [33,34] as

$$\frac{dn}{d(\Delta T)} = \frac{n_{max}}{\sqrt{2\pi}\Delta T_\sigma} exp\left[-\frac{(\Delta T - \Delta T_{max})^2}{2\Delta T_\sigma^2}\right] \tag{3}$$

where ΔT_{max}, ΔT_σ, and n_{max} are the mean undercooling, the standard deviation, and the maximum density of nuclei, respectively.

In casting, the total undercooling of the dendrite tip is generally the sum of four contributions [33,34], as follows

$$\Delta T = \Delta T_c + \Delta T_t + \Delta T_k + \Delta T_r \tag{4}$$

where ΔT_c, ΔT_t, ΔT_k, and ΔT_r are the undercoolings contributions associated with solute diffusion, thermal diffusion, attachment kinetics, and solid–liquid interface curvature, respectively.

They are the undercooling contributions associated with solute diffusion, thermal diffusion, attachment kinetics, and solid–liquid interface curvature. The last three contributions are small for the solidification process for making MMCs, and the solute undercooling predominated.

During solidification, the constitutional supercooling and kinetic undercooling affect the dendrite growth. In general, constitutional supercooling plays a decisive role in the growth of the dendrite tip. Thus, its growth kinetics can be predicted effectively using KGT (Kurz–Givoanola–Trivedi) model [34]. Hence, the growth rate formula for the dendrite tip can be expressed as [33,34]

$$\vartheta_{tip} = \alpha(\Delta T)^2 + \beta(\Delta T)^3 \tag{5}$$

where α and β are empirical constants.

2.2. *Material Properties*

The material properties of the fiber, the nickel coating, and the melt used in the analysis are presented in Table 1. Values of some primary parameters used in the microstructure simulation are shown in Table 2. (These parameters on microstructure growth are mainly taken from the ProCast manual.) Coefficients of the growth kinetics are calculated by the module in the ProCast2009 software. The calculated results are quite consistent with the values available in the literature.

Table 1. Material parameters gleaned from [35] and used in ProCast's CAFE simulation.

Property	Carbon Fiber	Nickel	Al-2014
Thermal conductivity (W/m C)	54	60.7	193
Specific heat capacity (J/kg K)	921	460	880
Density (kg/m^3)	1800	8880	2800
Thermal expansion coefficient (m/m °C)	-10^{-7}	13×10^{-6}	23×10^{-6}
Young's modulus (GPa)	217	207	71
Poisson's ratio	0.3	0.31	0.33

Table 2. The in-built parameters used in ProCast's CAFE simulation.

Property		Value
a2(First coefficient of the growth kinetics)		4.7×10^{-6}
a3(Second coefficient of the growth kinetics)		2.5×10^{-7}
Nucleation parameters in the bulk of the liquid (Gaussian distribution)	DTm (Average undercooling)	2.5
	DTs (Standard deviation)	1
	Nmax (Maximum number of nuclei)	7×10^{10}
Nucleation parameters at the surface (Gaussian distribution)	dTm (Average undercooling)	0.5
	dTs (Standard deviation)	0.1
	Gmax (Maximum number of nuclei)	5.0×10^{10}

2.3. *Initial and Boundary Conditions*

The adiabatic thermal boundary condition is applied to the outer surfaces of the model except at the bottom of the carbon fiber and coating. The carbon fiber's bottom surface (z = 0) is given a constant temperature of 25 °C. The initial temperature of the fiber and the aluminum melt is considered to be 639 °C. It is assumed that the melt–fiber interface is ideal in terms of contact conditions, while the inner surface of the carbon fiber and coating are in complete contact, and the formula for the coefficient of the interface between coating and melt can be expressed as

$$\lambda \frac{\partial T}{\partial n}\bigg|_{w} = h_i(T_{w1} - T_{w2}) \tag{6}$$

where $\lambda \frac{\partial T}{\partial n}\big|_{w}$ and h_i are the normal temperature-gradient-driven heat flux at the boundary (with λ being the thermal conductivity) and the boundary heat transfer coefficient, respectively. T_{w1} and T_{w2} represent the surface temperatures of the coating and melt, respectively.

For the thermal-stress analysis, a zero-displacement boundary condition is employed on the outer surface of the mesh model, which prevents deformation in the normal direction, but allows displacement in the tangential direction. Thermal stresses are calculated from the temperature field at any given time. The melt–fiber or melt–coating interfaces are assigned as nucleation sites for the microstructure simulation.

2.4. *Numerical Solution*

In this paper, the finite element method (FEM) is used for the numerical solution of the problem. The calculation domain should be simplified as a microunit (cylinder), which size is the one unit of height and one unit of diameter, as shown in Figure 1.

In the simulation coupling temperature and stress, though the temperature field is calculated at each time step, the coupled thermal stress field begins to be calculated when the solidification fraction of the melt reaches 50%. The iterative procedure is continued until the values at each node converge. The calculating procedure is stopped when the liquid metal is completely solidified.

In the solidification microstructure simulation, the cellular automaton (CA) finite element (FE) model is used to simulate the microstructure of Al alloy composites. In the simulation domain, the larger mesh is used to simulate the temperature and enthalpy defined at each node using the energy equation. Subsequently, the cell meshes with smaller size are used for microstructure calculations interactively by the CA method within the temperature range calculated by the FE method at each macrotime step. The nucleation and the dendritic-growth computations within the CA method are two significant components of the microstructure simulation described in Section 2.1.3.

The flow charts of the numerical solution procedure developed for this simulation are shown in Figure 2. The present study is proposed to compute changes in the temperature and thermal stress fields with time and to determine the microstructure evolution during the solidification process of the Al alloy considered. The thermal stress simulation and thermal grain structure simulation are calculated using the commercial software ProCast in our study. The solution domain is discretized into 174,661 nodes, the mesh grid size is about 2.5×10^{-3} mm, and the conservation equations are solved at each node of the elements. Such a procedure is repeated iteratively until convergence to the correct solution is obtained. The minimum time-step is about 0.001 s, and the convergence criteria for energy equation is 10^{-6}. During the presentation of the results, some variables involved in the governing equations are rendered dimensionless, including geometry, temperature, and time equations [17,30]:

$$\bar{r} = \frac{r}{R_a}, \ \bar{z} = \frac{z}{R_a}, \ \overline{R_f} = \frac{R_f}{R_a} \quad (7)$$

$$AR = \frac{L}{R_a}, \theta_m = \frac{T_m - T_0}{T_i - T_0}, F_o = \frac{t\alpha}{R_a^2}, \ \overline{HF} = \frac{HF}{-k_a(T_i - T_0)/R_a} \quad (8)$$

where r is the radial distance, k_a is the thermal conductivity of the alloy, and T_m, T_0, and T_i is the temperature of the alloy, the cooling temperature, and the initial temperature, respectively. The symbol t represents the time, and a is the thermal expansion coefficient. The symbol with a bar represents the dimensionless parameters. AR is the aspect ratio of the mold, the dimensionless temperature of the alloy, the dimensionless time, and the dimensionless heat flux.

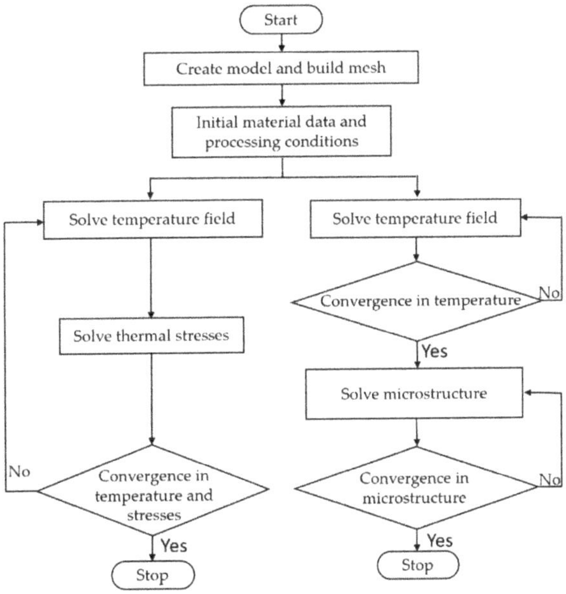

Figure 2. Flow chart of the numerical solution procedure adapted for modeling the solidification process of MMCs.

3. Results and Discussion

In the present study, the aluminum alloy melt has been pressurized and completely infiltrated into the mold packed with the fiber preform. The infiltration flow and the concomitant interface reactions are neglected. However, the effects of the cooling conditions and the coating on the temperature, thermal stress, and microstructure are considered in this analysis.

Figure 3 compares temporal changes in the temperature profile obtained from the axisymmetric model with and without nickel coating. The entire heat is extracted from the bottom of the extension fiber, so the low-temperature region is mainly concentrated in the lower part of the model. Meanwhile, the effect of the active cooling leads to a large temperature gradient near the fiber end. The nickel coating can act as a thermal barrier layer and make the melt's temperature gradient steeper near the coating–melt interface. As we shall see later, this effect will result in a finer grain structure using nickel coating (Figures 14 and 15).

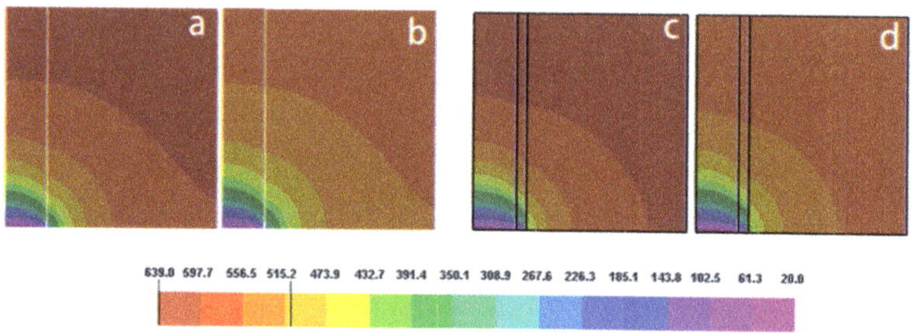

Figure 3. Temperature profile in axial section of the model without coating (**a**,**b**) and with coating (**c**,**d**) at different dimensionless times (**a**,**c**) Fo = 0.156, (**b**,**d**) Fo = 0.234.

The present simulation was verified with the results reported by Nguyen et al. [30] for the numerical simulation using a 2-D model. Figure 4 compares the temperature distribution for the same conditions. The results from the present 3-D simulation show reasonably good quantitative agreement with the results obtained by Nguyen. From this figure, it can also be seen that the temperature gradient of the melt decreases with increases in the axial and radial distances.

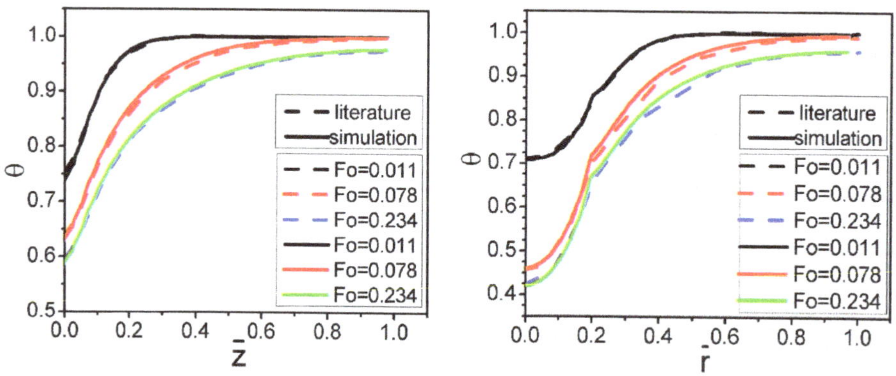

(**a**) (**b**)

Figure 4. Comparison of the results from the present simulation with that reported in the literature [30] on the evolution of temperature without the nickel coating: (**a**) $\bar{r}$ = 0.25, (**b**) $\bar{z}$ = 0.1.

Dimensionless heat flux as a function of radial location is studied for four axial-location cases and different time cases in Figure 5. It can be seen that the heat flux within the fiber changes little along the radial direction, but the heat flux in the melt declines sharply. There is a discontinuity in the heat flux profile due to a sharp change in the thermal properties across the fiber and melt interface. It is a potential site for stress concentration. Figure 5a also illustrates that the heat flux has an obvious decreasing trend away from the cooling end along the z-direction. Figure 5b shows the changes in the nondimensional heat flux along the axial (z) direction at the different time instances. In Figure 6, the heat flux obviously decreases along the axial direction as the radial distance increases from r = 0.19 to r = 0.26. However, there are differences in the two figures, because r = 0.19 is located inside the fiber and is near the cold fiber-end, while r = 0.26 is located in the melt. Inside the fiber, the decline in heat flux is much steeper than that in the melt, which may be attributed to differences in thermal conductivity in the two regions (Table 1).

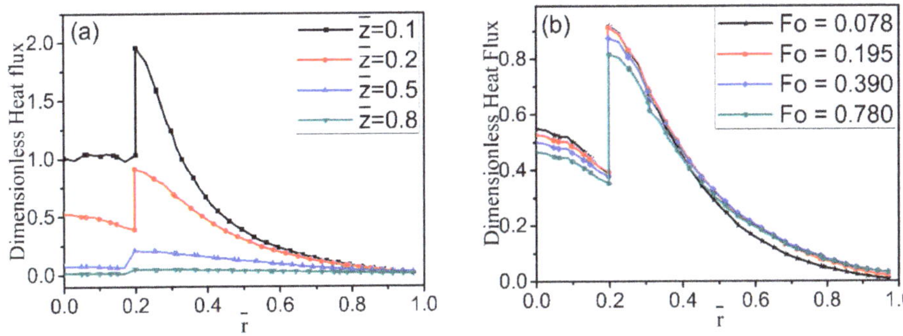

Figure 5. Effect of different distances on dimensionless heat flux without the nickel coating: (**a**) at Fo = 0.195; (**b**) at $\bar{z}$ = 0.2.

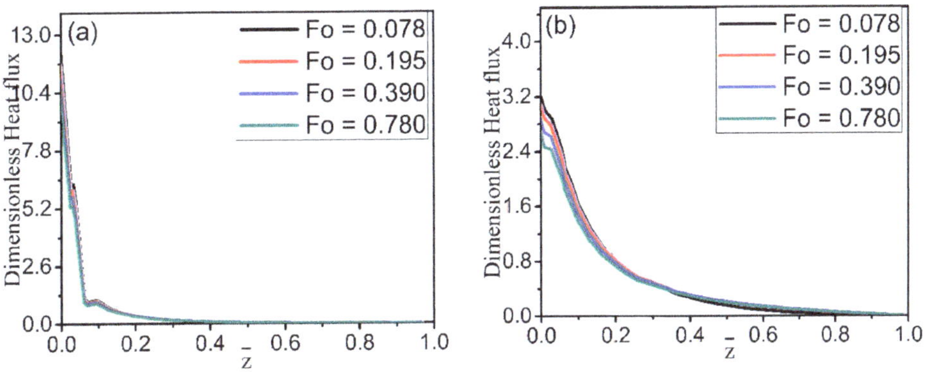

Figure 6. Changes in the dimensionless heat flux (without the nickel coating) along the fiber-axis direction at (**a**) $\bar{r}$ = 0.19, (**b**) $\bar{r}$ = 0.26.

Figure 7 compares the heat flux with and without nickel coating along the radial direction at the given z-direction location. It shows that the heat flux with nickel coating is less jagged than without nickel coating. The heat fluxes have a slight decline on the side of the fiber, then the expected increase near the interface, and then the smooth decrease on the melt side. Compared with the coating case, the increase in the heat flux is sharper near the interface for the no-coating case. This abrupt change is expected to produce a severe thermal stress concentration at the interface to result in the debonding of coating/fiber and alloy.

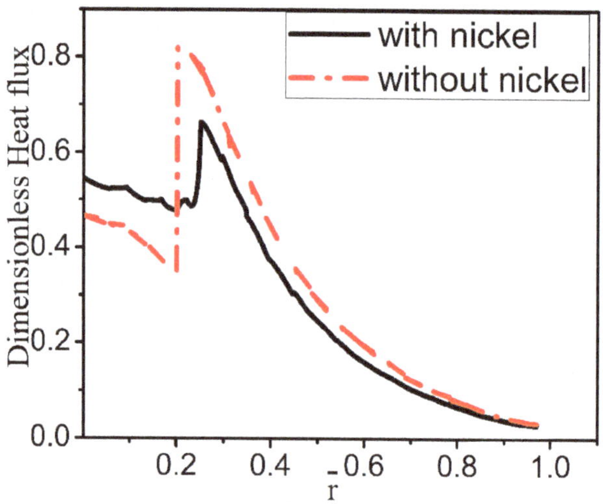

Figure 7. Dimensionless heat flux with and without nickel coating at $\bar{z} = 0.1$.

The thermal stress analysis is very effective for predicting stress concentrations, the resulting hot cracks, etc. The thermal stress is caused by temperature gradients in the fiber/alloy system. Residual stresses usually lead to thermal deformations in the material. Poor properties of metal matrix composites are often the result of undesirable heat distortions. Additionally, if the coating is considered in the model, the displacement caused by thermal deformation will become more complicated due to the different thermal expansion coefficients of the coating. Furthermore, nickel coating also has the function of transferring and bearing stresses.

The stress concentration exists at the interface due to higher von Mises stress. As shown in Figure 8, there is a sharp decrease in the stress on both sides of the fiber–melt interface. On the side of the fiber, the thermal stresses are relatively stable; however, it declines sharply on the melt side. It is also clear that the stress 'spikes' decrease in intensity as one moves away from the cooled part of the fiber at =0. The spikes are correlated with the temperature gradients shown in Figure 3.

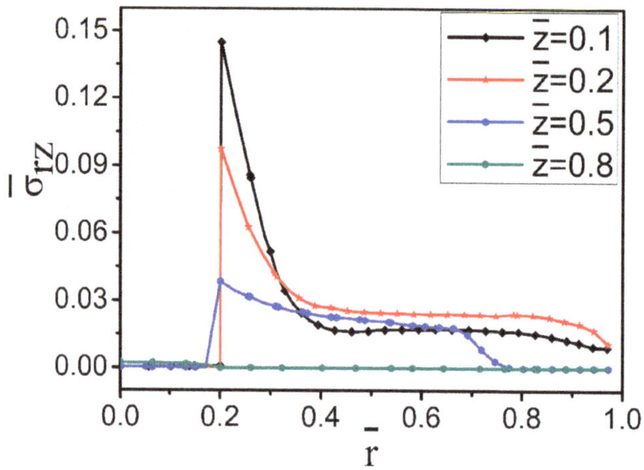

Figure 8. Variation in von Mises stress along the radial direction for the no-nickel-coating case at dimensionless time Fo = 0.195.

Figure 9 compares the resultant von Mises stresses with and without nickel coating. It can be seen that von Mises stress at the fiber–coating interface is much higher than at the fiber–melt interface without the coating. A close observation of this result reveals that nickel coating can cause the residual stress to increase significantly, and the stress in the fiber is higher than that in the melt. This higher stress may result in debonding of the coating. So, it is necessary to effectively control the cooling rate to protect against coating failure [36].

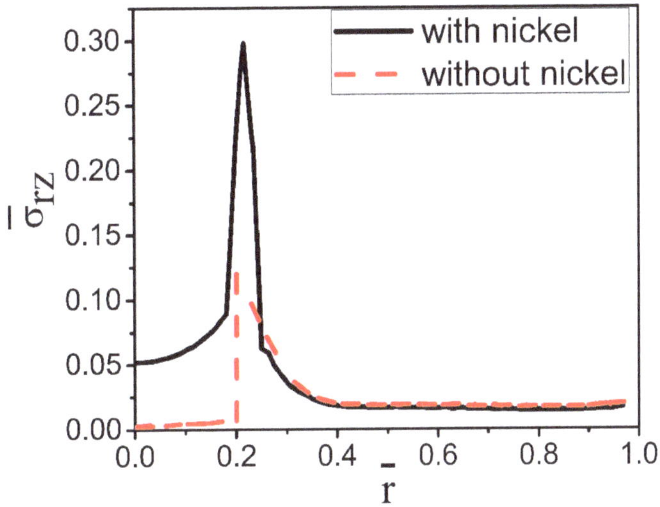

Figure 9. Radial changes in von Mises stress with and without nickel coating $\bar{z} = 0.1$.

Figures 10 and 11 illustrate the radial and axial deformation profiles at different cooling times. It can be seen that the extent of deformation, especially the axial one, in the fiber side with the nickel coating is far greater than that without the coating. However, the deformations are much smaller in the melt.

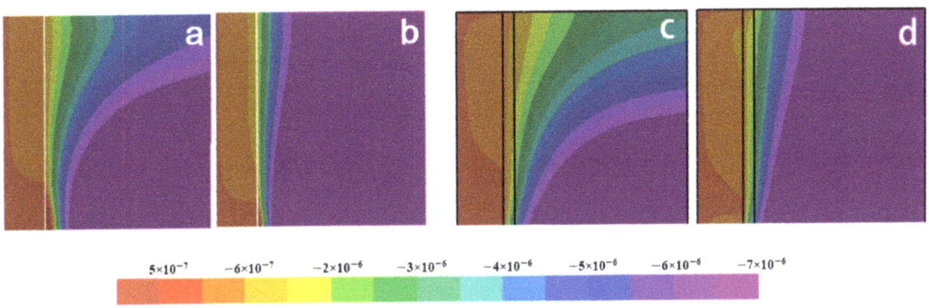

Figure 10. Distributions of radial deformation: (**a**) Fo = 0.156, without coating; (**b**) Fo = 0.234, without coating; (**c**) Fo = 0.156, with coating; (**d**) Fo = 0.234, with coating.

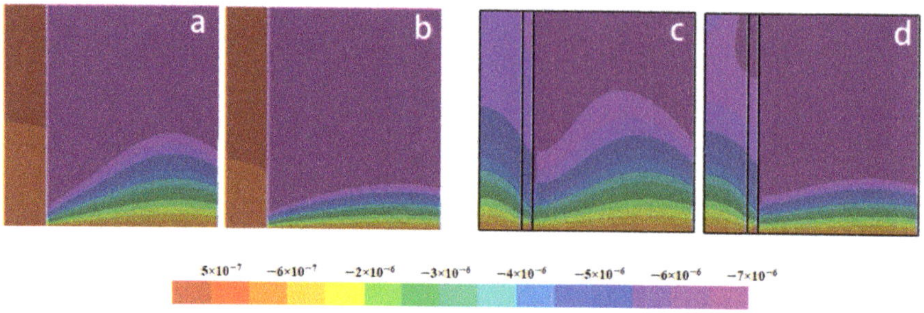

Figure 11. Distributions of axial deformation: (**a**) Fo = 0.156, without coating; (**b**) Fo = 0.234, without coating; (**c**) Fo = 0.156, with coating; (**d**) Fo = 0.234, with coating.

Figure 12 describes in detail comparison between the distributions of deformation with and without nickel coating. It can be found that the deformation displacement of the model with nickel coating is smaller than that without coating in the radial direction at $\bar{z} = 0.2$. At the same time, there is little change in displacement at the fiber side. Figure 12b, $\bar{r} = 0.19$ reveals a completely different deformation of the model with and without nickel coating. The main reason is the different thermal expansion coefficient values for fiber, coating, and melt. The carbon fiber shows volume expansion with decreasing temperature, while the volume shrinks due to the positive thermal expansion coefficient of nickel coating and melt. In addition, the absolute value of the thermal expansion coefficient of the melt and the nickel coating is far greater than that of the fiber. So, the expansion process of the fiber is blocked and forced to move in the opposite direction during the cooling process. Figure 12b, $\bar{r} = 0.4$ presents that the deformation displacement of the model with nickel coating is smaller than without coating.

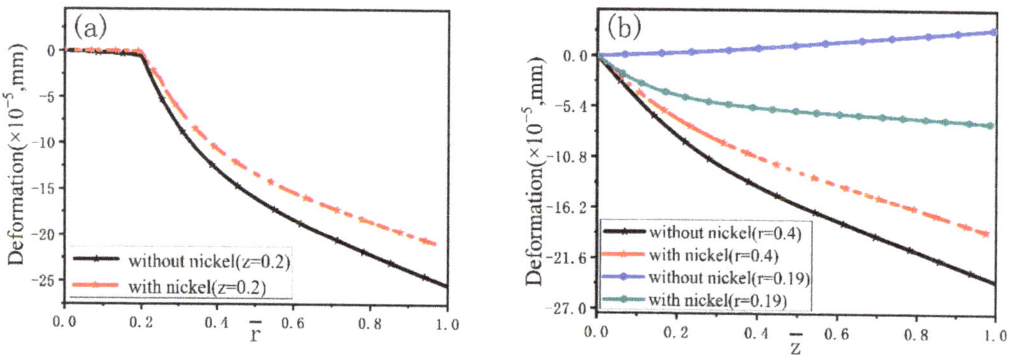

Figure 12. Deformation with and without nickel coating: (**a**) at $\bar{z} = 0.2$, (**b**) at $\bar{r} = 0.19$ and $\bar{r} = 0.4$.

The nucleation and growth of grains during solidification is a key factor in determining the grain size and achieving the desirable properties of MMCs. It is not only related to the material itself but also closely connected with the cooling rate. Experimental investigations show that the grain size of the composites is often smaller than that of the unreinforced alloy under identical solidification conditions. Solute diffusion is impeded during growth due to the barrier effects of the reinforcement. Therefore, the delayed growth from the melt gives additional time for the formation of nuclei, which can yield a refined structure [27,28]. Microstructures in fiber-reinforced MMCs can be modulated in a predetermined manner by controlling interfiber spacing and cooling rate [28]. In the present study, the effects of cooling conditions and coating on grain microstructure were studied using the CAFÉ model. The cooling condition was controlled by varying the degree of cooling of the ends of fibers with water cooling: 25 °C and dry ice–acetone mixture: −78 °C.

Figure 13 shows a series of axial and radial slices of grain microstructure predicted. Figures 13–15 present the simulation results of the grain microstructure profile of axial directions with and without coating under different cooling conditions. The higher temperature gradient is obviously affecting the microstructure close to the cold end. The grain growth near the nucleation surface is restrained and forms finer equiaxial crystals due to chilling. However, grains tend to form coarse columnar crystals away from the chilling layer, as shown in Figure 14(a–c,a1–c1) and Figure 15(a–c,a1–c1). The microstructure away from the fiber tends to form coarse equiaxed grains due to the lower thermal gradient away from the cold end, as shown in Figure 14(d,e,d1,e1) and Figure 15(d,e,d1,e1). Compared to the simulated results with and without nickel coating, it can be found from Figure 14(e,e1) and Figure 15(e,e1) that the number of grains increases significantly in the presence of the nickel coating away from the fiber. The size of grains also reduces relatively, as observed in Figures 14 and 15. The main feature is that the nickel coating can act as a thermal resistance layer, making the temperature profiles near interfaces uniform and further reducing the temperature gradient, obtaining the fine grain structure. Similarly, this argument can be demonstrated in Figure 16. A close observation of Figure 16 reveals that as the location of the slice is far away from the carbon fiber (from (a) to (e)), the number of grains firstly increases and then decreases, the mean surface of grain firstly decreases and then increases. Moreover, it can be seen that there is a small difference in the number of grains and the average grain area at different cooling temperatures.

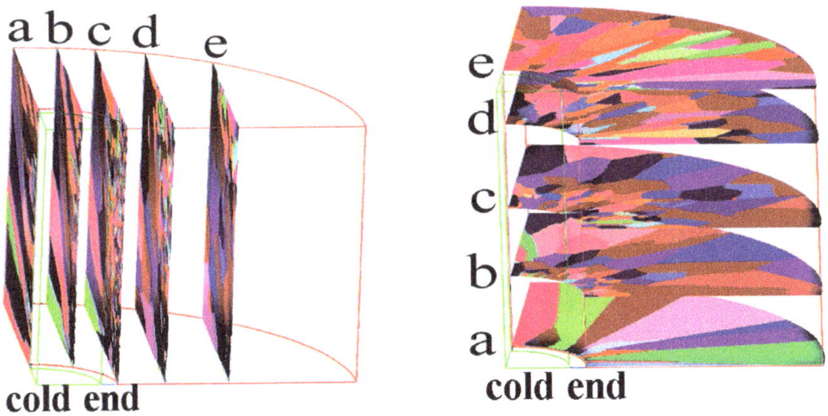

(a) (b)

Figure 13. Microstructure predicted at the different sections: (**a**) axial sections, (**b**) radial sections.

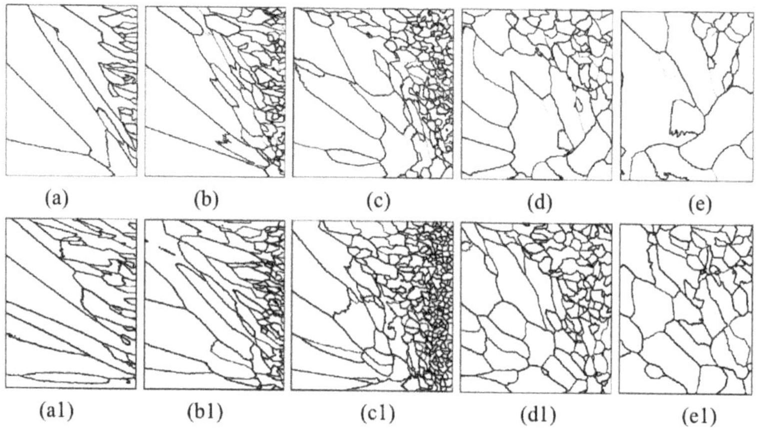

Figure 14. Microstructure morphology predicted of different axial sections with the cooling temperature of 25 °C (**a–e**) without nickel coating; (**a1–e1**) with nickel coating.

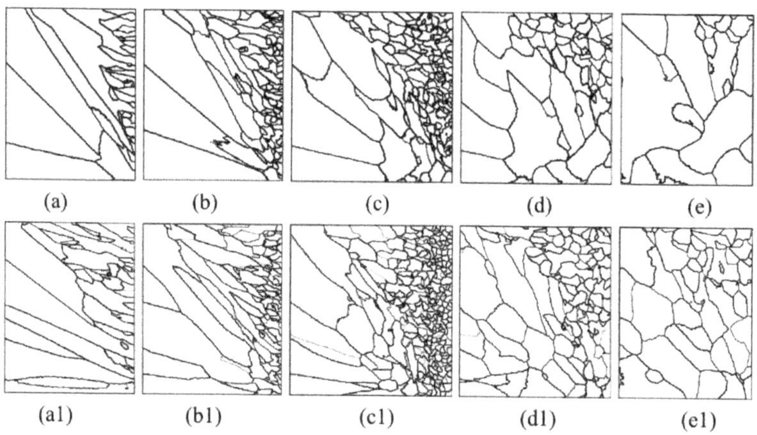

Figure 15. Microstructure morphology predicted of different axial sections with the cooling temperature of −78 °C (**a–e**) without nickel coating; (**a1–e1**) with nickel coating.

The microstructure morphology of the different radial sections is presented in Figures 17 and 18. When slices are away from the cold end (from slice (a) to slice (e), as shown in the right picture of Figure 13), the grains will gradually become equiaxed grains at the edge of the slice. On the contrary, the grains will develop columnar crystals at the center of the section (near the carbon fiber). The reason for this situation is that the crystal growth direction is always in the opposite direction of heat flow.

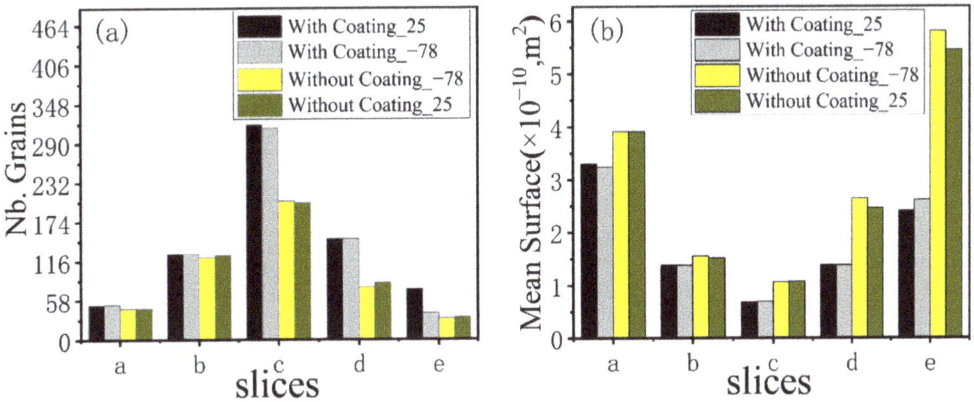

Figure 16. Number grains (**a**) and mean surface (**b**) of different axial sections.

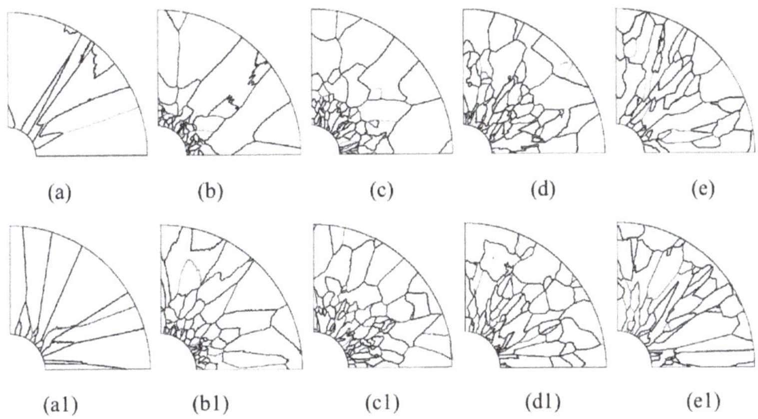

Figure 17. Microstructure morphology predicted of different radial sections with the cooling temperature of 25 °C (**a–e**) without nickel coating; (**a1–e1**) with nickel coating.

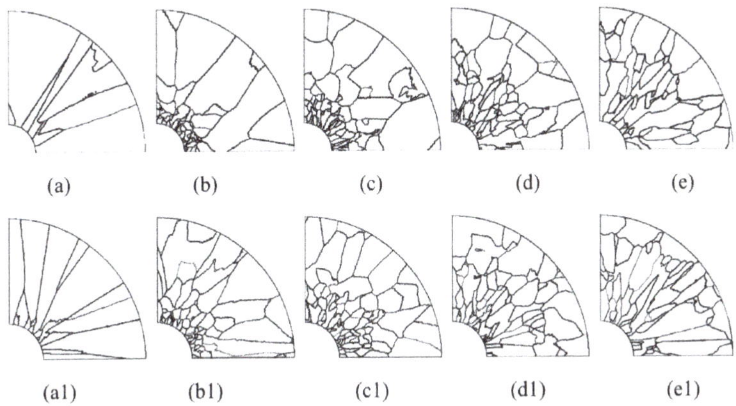

Figure 18. Microstructure morphology predicted of different radial sections with the cooling temperature of −78 °C (**a–e**) without nickel coating; (**a1–e1**) with nickel coating.

Comparing the simulated microstructure of radial section with and without coating shows that the number of grains increases significantly, and the grains' mean surface slightly decreases when using the nickel coating, which can be proved via Figure 19. Furthermore, Figure 19 also shows that the number of grains increases firstly and then decreases, and the mean surface presents the opposite trend with the increasing distance from the cold end. In conclusion, the microstructure is obviously refined, which is beneficial for improving the properties of metal matrix composites when the fiber is wrapped with nickel coating and the extended fiber cooling method.

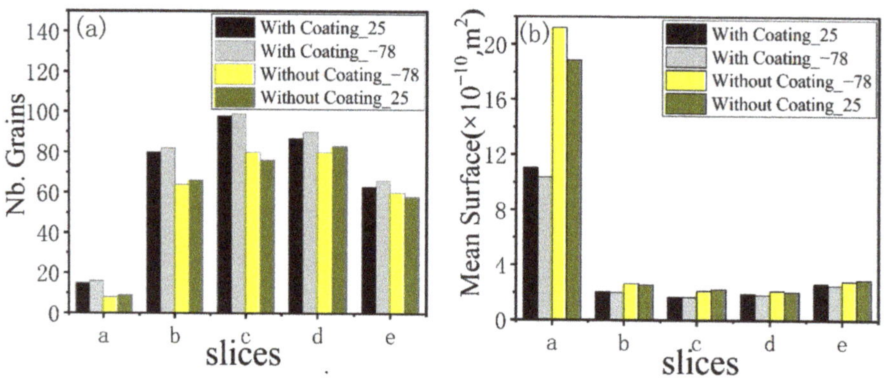

Figure 19. Number grains (**a**) and mean surface (**b**) of different radial sections.

An earlier study by Rohatgi [37] and Lee [27] revealed that cooling the extended ends of the reinforcement results in finer microstructures in the matrix and changes the nature of the interface. Figure 20 demonstrates the effect of fiber cooling on the matrix microstructure of an aluminum/carbon fiber composite, in which the carbon fibers were chilled outside of the mold [27]. In this case, very fine-sized grains were in contact with the surfaces of graphite fibers. The same results also can be observed from Figure 13, Figure 17, and Figure 18. The predicted results of solidification microstructure agreed well with these experimental results. So, it is demonstrated that the cooling of the extent of the fibers during making MMCs can be useful for producing the finer matrix microstructures around fibers and improving the properties of MMCs. However, suitable cooling conditions need to be designed and optimized in future works.

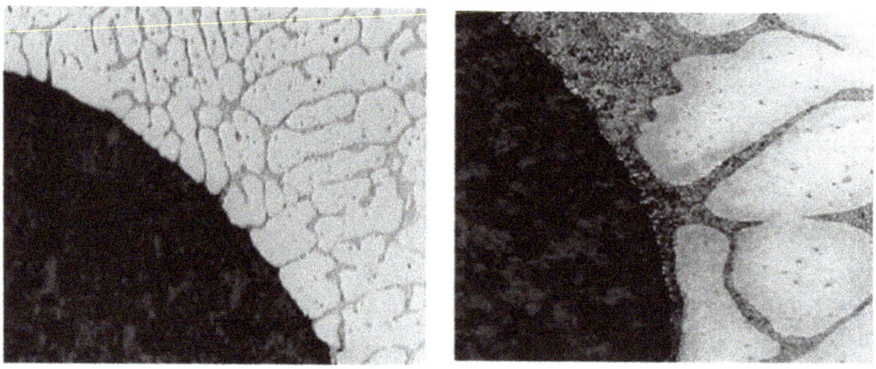

(**a**) (**b**)

Figure 20. Solidifying microstructure of thermally managed Al-9% Cu alloy composite (**a**) with external cooling of graphite rod extending out of the melt and (**b**) without external cooling of the graphite rod [27].

4. Conclusions

This paper mainly focuses on the numerical simulation of the temperature field, thermal stress, and microstructure of the fiber-reinforced metal matrix composites. Two physical models with and without nickel coating are analyzed in detail. At the same time, the effects of cooling rate on the resultant microstructure are also investigated by changing the fiber end cooling temperature. The following important conclusions are drawn from this study:

1. Based on a modified infiltration process by Nguyen et al., the effect of active cooling conditions on temperature distribution was simulated. The predicted results of temperature evolution agreed well with the reported results.
2. The distribution of heat flux has a significant influence on the microstructure and thermal stress. The heat flux trend is gradually evolving from the top of the model to the bottom of the fiber due to active cooling through carbon fiber. On the side of the fiber, the heat flux changes smoothly, while it varies drastically at the melt side. Comparing analysis results of the heat flux with and without nickel coating reveals that it is smoother and smaller in the Ni-coating model, which is favorable for preventing debonding at the interface of coating/fiber and alloy and obtaining the finer grains.
3. The predicted results of the thermal stress show that there is high thermal stress on the interfaces of fiber–coating, coating–melt, and fiber–melt. These places tend to cause stress concentration. On the one hand, it is easy to generate microcracks in these locations, resulting in interface failure; on the other hand, it tends to lead to debonding of the coating.
4. The formation and growth of grains are closely related to the temperature field. The heat is only dissipated from the bottom of the fiber. Therefore, the dendrites obliquely grew along with the model from the lower part of the fiber. The number of grains near the nucleation is more than that of the other places due to the effect of chilling. We also can see that the microstructure is significantly refined, and then the properties of metal matrix composites can be improved when the fiber is wrapped by nickel coating.

Author Contributions: Conceptualization, C.X. and B.W.; methodology, B.W.; software, C.X. and Q.W.; validation, R.E., K.M.P. and B.W., formal analysis, C.X. and Q.W.; investigation, C.X., Q.W. and B.W.; resources, B.W.; data curation, B.W.; writing—original draft preparation, C.X. and B.W.; writing—review and editing, C.X., R.E., K.M.P. and B.W.; visualization, B.W.; supervision, B.W.; project administration, B.W.; funding acquisition, B.W. All authors have read and agreed to the published version of the manuscript.

Funding: The authors gratefully express their appreciation to the Innovation Program of the Shanghai Municipal Education Commission (No. 2019-01-07-00-09-E00024). This work was supported by the Independent Research and Development Project of State Key Laboratory of Advanced Special Steel, Shanghai Key Laboratory of Advanced Ferrometallurgy, Shanghai University (SKLASS 2021-Z02), and the Science and Technology Commission of Shanghai Municipality (No. 19DZ2270200, 20511107700).

Institutional Review Board Statement: Not applicable.

Informed Consent Statement: Not applicable.

Data Availability Statement: Not applicable.

Conflicts of Interest: The authors declare no conflict of interest.

References

1. Garg, P.; Jamwal, A.; Kumar, D. Advance research progresses in aluminium matrix composites: Manufacturing & applications. *J. Mater. Res. Technol.* **2019**, *8*, 4924–4939.
2. Huda, M.D.; Hashmi, M.S.J.; El-Baradie, M.A. MMCs: Materials, Manufacturing and Mechanical Properties. *Key Eng. Mater.* **1995**, *104–107*, 37–64. [CrossRef]
3. Hashim, J.; Looney, L.; Hashmi, M.S.J. Particle distribution in cast metal matrix composites—Part I. *J. Mater. Process. Technol.* **2002**, *123*, 251–257. [CrossRef]

4. Rohatgi, P.K.; Kumar, P.A.; Chelliah, N.M.; Rajan, T.P.D. Solidification Processing of Cast Metal Matrix Composites Over the Last 50 Years and Opportunities for the Future. *JOM* **2020**, *72*, 2912–2926. [CrossRef]
5. Surappa, M.K. Aluminium matrix composites: Challenges and opportunities. *Sadhana* **2003**, *28*, 319–334. [CrossRef]
6. Wielage, B.; Wank, A.; Wilden, J. Production of Composites or Bonding of Material by Thermal Coating Processes. In *Metal Matrix Composites*; John Wiley & Sons, Ltd.: New York, NY, USA, 2006; pp. 111–146.
7. Buschmann, R. Preforms for the Reinforcement of Light Metals—Manufacture, Applications and Potential. In *Metal Matrix Composites*; John Wiley & Sons, Ltd.: New York, NY, USA, 2006; pp. 77–94.
8. Deng, K.; Shi, J.; Wang, C. Microstructure and strengthening mechanism of bimodal size particle reinforced magnesium matrix composite. *Compos. Part A Appl. Sci. Manuf.* **2012**, *43*, 1280–1284. [CrossRef]
9. Demir, A.; Altinkok, N. Effect of gas pressure infiltration on microstructure and bending strength of porous Al_2O_3/SiC-reinforced aluminium matrix composites. *Compos. Sci. Technol.* **2004**, *64*, 2067–2074. [CrossRef]
10. Etemadi, R.; Wang, B.; Pillai, K.M. Pressure infiltration processes to synthesize metal matrix composites—A review of metal matrix composites, the technology and process simulation. *Mater. Manuf. Processes* **2018**, *33*, 1261–1290. [CrossRef]
11. Mortensen, A. Melt infiltration of metal matrix composites. *Compr. Compos. Mater.* **2000**, *3*, 521–524.
12. Al-Mosawi, B.T.S.; Wexler, D.; Calka, A. Characterization and Properties of Aluminium Reinforced Milled Carbon Fibres Composites Synthesized by Uniball Milling and Uniaxial Hot Pressing. *Met. Mater. Int.* **2021**, *27*, 3617–3640. [CrossRef]
13. Hashim, J.; Looney, L.; Hashmi, M.S.J. The wettability of SiC particles by molten aluminium alloy. *J. Mater. Processi. Technol.* **2001**, *119*, 324–328. [CrossRef]
14. Malaki, M.; Tehrani, A.F.; Niroumand, B.; Gupta, M. Wettability in Metal Matrix Composites. *Metals* **2021**, *11*, 1034. [CrossRef]
15. Shirvanimoghaddam, K.; Hamim, S.U.; Akbari, M.K.; Fakhrhoseini, S.M.; Khayyam, H.; Pakseresht, A.H.; Ghasali, E.; Zabet, M.; Munir, K.S.; Jia, S.; et al. Carbon fiber reinforced metal matrix composites: Fabrication processes and properties. *Compos. Part A Appl. Sci. Manuf.* **2017**, *92*, 70–96. [CrossRef]
16. Vidal-Setif, M.; Lancin, M.; Marhic, C. On the role of brittle interfacial phases on the mechanical properties of carbon fibre reinforced Al-based matrix composites. *Mater. Sci. Eng. A* **1999**, *272*, 321–333. [CrossRef]
17. Gupta, N.; Nguyen, N.Q.; Rohatgi, P.K. Analysis of active cooling through nickel coated carbon fibers in the solidification processing of aluminum matrix composites. *Compos. Part B Eng.* **2011**, *42*, 916–925. [CrossRef]
18. Daoud, A. Microstructure and tensile properties of 2014 Al alloy reinforced with continuous carbon fibers manufactured by gas pressure infiltration. *Mater. Sci. Eng. A* **2005**, *391*, 114–120. [CrossRef]
19. Urena, A.; Rams, J.; Escalera, M.; Sanchez, M. Characterization of interfacial mechanical properties in carbon fiber/aluminium matrix composites by the nanoindentation technique. *Compos. Sci. Technol.* **2005**, *65*, 2025–2038. [CrossRef]
20. Zhang, G.D.; Chen, R. Effect of the interfacial bonding strength on the mechanical properties of metal matrix composites. *Compos. Interfaces* **1993**, *1*, 337–355. [CrossRef]
21. Liu, Z.; Mang, X.; Chai, L.; Chen, Y. Interface study of carbon fibre reinforced Al–Cu composites. *J. Alloy. Compd.* **2010**, *504*, S512–S514. [CrossRef]
22. Mortensen, A.; Jin, I. Solidification processing of metal matrix composites. *Int. Mater. Rev.* **1992**, *37*, 101–128. [CrossRef]
23. Chang, C.-Y. Numerical simulation of the pressure infiltration of fibrous preforms during MMC processing. *Adv. Compos. Mater.* **2006**, *15*, 287–300. [CrossRef]
24. Moussa, B.; E Simpson, J.; Garimella, S.V. Concentration fields in the solidification processing of metal matrix composites. *Int. J. Heat Mass Transf.* **2002**, *45*, 4251–4266. [CrossRef]
25. Hanumanth, G.S.; Irons, G.A. Solidification of particle-reinforced metal-matrix composites. *Metall. Mater. Trans. B* **1996**, *27*, 663–671. [CrossRef]
26. Kosti, S.; Pathak, P. Genetic Algorithm based Finite Difference Simulation of Solidification Process for MMC's. *Mater. Today Proc.* **2018**, *5*, 8271–8279. [CrossRef]
27. Lee, E.-K.; Amano, R.S.; Rohatgi, P.K. Metal matrix composite solidification in the presence of cooled fibers: Numerical simulation and experimental observation. *Heat Mass Transf.* **2007**, *43*, 741–748. [CrossRef]
28. Yang, M.; Wang, L.; Yan, W. Phase-field modeling of grain evolution in additive manufacturing with addition of reinforcing particles. *Addit. Manuf.* **2021**, *47*, 102286. [CrossRef]
29. Lelito, J.; Zak, P.L.; Shirzadi, A.A.; Greer, A.L.; Krajewski, W.K.; Suchy, J.S.; Haberl, K.; Schumacher, P. Effect of SiC reinforcement particles on the grain density in a magnesium-based metal–matrix composite: Modelling and experiment. *Acta Mater.* **2012**, *60*, 2950–2958. [CrossRef]
30. Nguyen, N.Q.; Peterson, S.D.; Gupta, N.; Rohatgi, P.K. Modeling the effect of active fiber cooling on the microstructure of fiber-reinforced metal matrix composites. *Metall. Mater. Trans. A* **2009**, *40*, 1911–1922. [CrossRef]
31. Rohatgi, P.; Asthana, R. Solidification science in cast MMCs: The influence of Merton Flemings. *JOM* **2001**, *53*, 9–13. [CrossRef]
32. Rohatgi, P.K.; Tiwari, V.; Gupta, N. Squeeze infiltration processing of nickel coated carbon fiber reinforced Al-2014 composite. *J. Mater. Sci.* **2006**, *41*, 7232–7239. [CrossRef]
33. Li, X.-M.; Zhang, J.-Y.; Wang, B.; Ren, Z.-M.; Zhou, G.-Z. Simulation of stray grain formation during unidirectional solidification of IN738LC superalloy. *J. Cent. South Univ. Technol.* **2011**, *18*, 23–28. [CrossRef]
34. Kurz, W.; Giovanola, B.; Trivedi, R. Theory of microstructural development during rapid solidification. *Acta Metall.* **1986**, *34*, 823–830. [CrossRef]

35. Seong, H.; Lopez, H.; Rohatgi, P. Microsegregation during solidification of graphitic fiber-reinforced aluminum alloys under external heat sinks. *Metall. Mater. Trans. A* **2007**, *38*, 138–149. [CrossRef]
36. Vajari, D.A.; Sørensen, B.F.; Legarth, B.N. Effect of fiber positioning on mixed-mode fracture of interfacial debonding in composites. *Int. J. Solids Struct.* **2015**, *53*, 58–69. [CrossRef]
37. Rohatgi, P.K.; Narendranath, C.S.; Wang, D. *Temperature Differences between Model Cylindrical Rods and Alloys Solidifying around These Rods. Microstructure Formation during Solidification of Metal Matrix Composites*; The Minerals, Metals & Materials Society: Warrendale, PA, USA, 1993; pp. 149–160.

MDPI
St. Alban-Anlage 66
4052 Basel
Switzerland
www.mdpi.com

Materials Editorial Office
E-mail: materials@mdpi.com
www.mdpi.com/journal/materials

www.ingramcontent.com/pod-product-compliance
Lightning Source LLC
Chambersburg PA
CBHW061620210926
51850CB00009B/2381
9783725808564